Jack Xin

An Introduction to Fronts in Random Media

 Springer

Jack Xin
Department of Mathematics
University of California, Irvine
Irvine, CA 92697
USA
jxin@math.uci.edu

Editors:

S.S. Antman
Department of Mathematics
and
Institute for Physical Science
   and Technology
University of Maryland
College Park
MD 20742-4015
USA
ssa@math.umd.edu

J.E. Marsden
Control and Dynamical
   System, 107-81
California Institute
   of Technology
Pasadena, CA 91125
USA
marsden@cds.caltech.edu

L. Sirovich
Laboratory of Applied
   Mathemaics
Department of
   Bio-Mathematical Sciences
Mount Sinai School of Medicine
New York, NY 10029-6574
USA
Lawrence.Sirovich@mssm.edu

ISBN 978-0-387-87682-5   e-ISBN 978-0-387-87683-2
DOI 10.1007/978-0-387-87683-2
Springer Dordrecht Heidelberg London New York

Library of Congress Control Number: 2009926482

Mathematics Subject Classification (2000): 60H15, 60H30, 76M30, 76M50, 76M45

Printed on acid-free paper

Springer is part of Springer Science+Business Media (www.springer.com)

Dedicated with love to my wife, Lily,
my sons, Spencer and Grant,
and my parents, Dingding and Ningyuan

Preface

This book aims to give a user-friendly tutorial of an interdisciplinary research topic (fronts or interfaces in random media) to senior undergraduates and beginning graduate students with basic knowledge of partial differential equations (PDE) and probability. The approach taken is semiformal, using elementary methods to introduce ideas and motivate results as much as possible, then outlining how to pursue rigorous theorems, with details to be found in the references section.

Since the topic concerns both differential equations and probability, and probability is traditionally a quite technical subject with a heavy measure-theoretic component, the book strives to develop a simplistic approach so that students can grasp the essentials of fronts and random media and their applications in a self-contained tutorial.

The book introduces three fundamental PDEs (the Burgers equation, Hamilton–Jacobi equations, and reaction–diffusion equations), analysis of their formulas and front solutions, and related stochastic processes. It builds up tools gradually, so that students are brought to the frontiers of research at a steady pace.

A moderate number of exercises are provided to consolidate the concepts and ideas. The main methods are representation formulas of solutions, Laplace methods, homogenization, ergodic theory, central limit theorems, large-deviation principles, variational principles, maximum principles, and Harnack inequalities, among others. These methods are normally covered in separate books on either differential equations or probability. It is my hope that this tutorial will help to illustrate how to combine these tools in solving concrete problems.

The three basic equations go from their constant-coefficient forms, well studied in graduate textbooks, to their full stochastic glory with space–time-dependent random coefficients. However, they are all connected to Hamilton–Jacobi equations. The reaction–diffusion equations are classified. The KPP (Kolmogorov–Petrovsky–Piskunov) fronts are discussed in detail because of their connections with the Hamiltonian dynamics in classical mechanics and their elegant analysis. The recent mathematical advance in solving a long-standing turbulent combustion problem is presented for KPP fronts. The non-KPP reaction–diffusion fronts are associated with a Hamiltonian resembling that in special relativistic mechanics. The mechanical connections of reaction–diffusion fronts and KPP solution methods in the spirit of Lagrangian and Eulerian perspectives are explored.

The scope of the book goes from exact solutions of scalar deterministic PDEs (Burgers, Hamilton–Jacobi, reaction–diffusion equations) to the asymptotic solutions of their stochastic counterparts. The reader will come to appreciate new random phenomena step by step, and learn that exact solutions are harder and harder to come by, while asymptotic ones are more accessible.

The first chapter of the book discusses fronts in homogeneous media, and the second chapter is on fronts in periodic media. These two chapters serve as an introduction to differential equations, their front solutions and representation formulas, and homogenization methods.

The last three chapters introduce stochastic equations, representation formulas and asymptotics, stochastic homogenization, variational methods, and large-deviation methods for analyzing random fronts.

The first three chapters are adaptations of the author's 2000 SIAM review article with the inclusion of new results. The remaining two chapters are based on recent results on stochastic homogenization of Hamilton–Jacobi equations and KPP fronts in random media.

Acknowledgments

I would like to thank George Papanicolaou for his encouragement and interest in my work on random fronts over the years. I am grateful to James Nolen and Janek Wehr for many helpful conversations during the preparation of the book. I am thankful for the support of the National Science Foundation and the constructive comments of the anonymous referees.

Jack Xin
Irvine, California
February, 2009

Contents

Chapter 1
Fronts in Homogeneous Media

Front propagation and interface motion occur in many scientific areas such as chemical kinetics, combustion, biology, transport in porous media, and industrial deposition processes. In spite of these different applications, the basic phenomena can all be modeled using nonlinear partial differential equations or systems of such equations. Since the pioneering work of Kolmogorov, Petrovsky and Piskunov (KPP) [134] and Fisher [91] in 1937 on traveling fronts of the reaction–diffusion equations, the field has gone through enormous growth and development. However, studies of fronts in heterogeneous media have been more recent. Heterogeneities are always present in natural environments, such as fluid convection effects in combustion (wind factor in spreading of forest fires), inhomogeneous porous structures in transport of solutes, noise effects in biology, and deposition processes.

It is both a fundamental and a practical problem to understand how heterogeneities influence the characteristics of front propagation such as front speeds, front profiles, and front locations. Our goal here is to give a tutorial of recent results on front propagation in heterogeneous (especially random) media in a coherent and motivating manner. It is not the intention of the book to give a complete survey, and so the cited references will cover only a portion of the vast literature.

Depending on the applications, three prototype equations appear, they are conservation laws, Hamilton–Jacobi equations (HJ), and reaction–diffusion (RD) equations. A class of problems of fronts in heterogeneous media arises in solute transport through porous media, an important subject in groundwater and environmental science. When solutes (ions) migrate inside porous media, some of them tend to attach onto the surface of minerals or colloids due to the existence of nonneutralized electric charges at the surface or inside these minerals [158]. This surface effect is called adsorption, which often creates a retardation on the movement of solute substance. The transport equation for the concentration of a one-species solute is based on the conservation of mass. When the adsorption reaches equilibrium, which often happens in a much shorter time than the time scale of solute migration, one arrives at the following conservation law [39, 71] for the solute concentration C:

$$(\omega C + \rho \psi(C))_t = \nabla \cdot (\theta D \nabla C - \mathbf{v} C), \qquad (1.1)$$

J. Xin, *An Introduction to Fronts in Random Media*, Surveys and Tutorials in the Applied
Mathematical Sciences 5, DOI: 10.1007/978-0-387-87683-2_1,
© Springer Science + Business Media, LLC 2009

where D is the pore-scale dispersion (viscosity) matrix, $\mathbf{v}$ is the incompressible water-flow velocity, ω is the total porosity, and $\rho = (1-\omega)\rho_s$, ρ_s are the densities of solid particles (minerals and colloids). The function $\psi = \psi(C)$ is called the sorption isotherm. For example, the Freundlich isotherm is of the form $\psi(C) = \kappa C^p$, $p \in (0,1)$, where κ represents the spatial distribution of sorption sites. Due to the heterogeneous nature of the porous media, both $\mathbf{v}$ and κ are functions of the spatial variable x. The lack of detailed field information (uncertainties) naturally leads to statistical modeling. Both $\mathbf{v}$ and k are treated as random processes [253, 38, 39, 197]. If $D = 0$ (inviscid regime), a change of variable $u = \omega C + \rho \psi(C)$ converts (1.1) into the standard form

$$u_t + \nabla \cdot \mathbf{f}(x,u) = 0, \tag{1.2}$$

where $\mathbf{f}$ is a vector stochastic flux function. The celebrated Burgers equation results if the flux function is scalar and equal to $u^2/2$. Since solutions of the inviscid equation (1.2) may become discontinuous [141], we shall view them as weak solutions obtained from zero viscosity limit of viscous solutions ($D \downarrow 0$). In this tutorial, sometimes the viscous version of the conservation laws will be analyzed for ease of representation of fronts and connections with physical modeling and reaction–diffusion equations. Many other equations of the form (1.2) arise in hydrology and multiphase flows, for example the Richards equation [195, 196, 88] and the Buckley–Leverett equation [118], to name just a few.

In manufacturing of nanomaterials and computer chips, atoms are deposited onto a substrate; then they may wander, diffuse, and stick to form a complicated landscape (interface) in the growth process of thin films. In continuum modeling of interface growth, a quadratic HJ equation, known as the Kardar, Parisi, and Zhang (KPZ) equation [13, Chapter 6],

$$h_t = \nu \nabla^2 h + \frac{\lambda}{2}|\nabla h|^2 + \eta(x,t), \tag{1.3}$$

describes the evolution of the interface height, where $\nu > 0$ is the diffusion constant, λ is a growth constant (positive if adding material to the interface), and η is a space–time uncorrelated (white) noise reflecting the random fluctuations in the deposition process. The gradient of h satisfies a viscous stochastic Burgers equation. In modeling of turbulent combustion of premixed flames, another HJ equation, known as the G-equation, is proposed [156, 238] and extensively studied for thin flame fronts in advecting fluid [194, 126, 127]:

$$G_t + \mathbf{v} \cdot \nabla_x G = s_L |\nabla_x G|, \tag{1.4}$$

where the scalar function G is the level set function of the flame front, $\mathbf{v}$ is a stochastic flow field, and s_L is the (laminar) flame speed in the absence of the flow. The level set $G(x,t) = G_0$, represents a moving interface; $G > G_0$ is the burnt (hot) region, $G < G_0$ is the unburnt (cold) region. Equation (1.4) amounts to saying that the normal velocity of the front is equal to $s_L + \mathbf{v} \cdot \mathbf{n}$, where $\mathbf{n}$ is the normal direction pointing to the unburnt region.

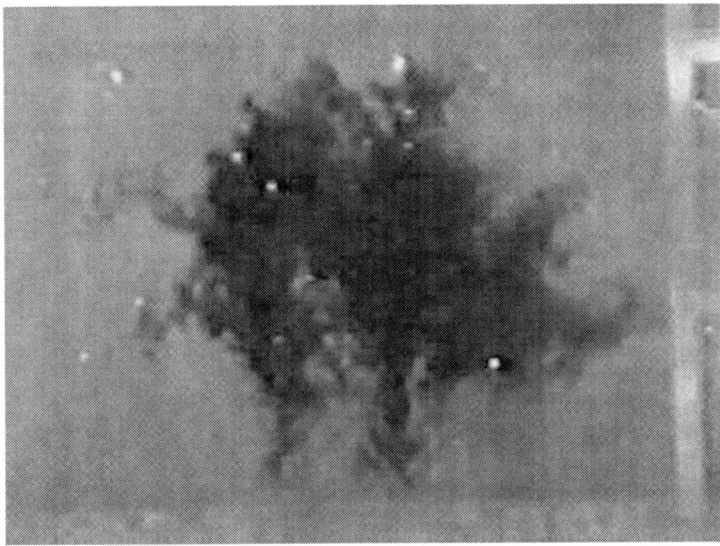

Figure 1.1 Example of a reactive front in a random medium. Planar laser-induced fluorescence light sheet image of an experimental arsenous-acid/iodate autocatalytic expanding front in random capillary wave flow of broad scales. Field view 14 cm $\times$ 14 cm, with the ratio U of root mean square (rms) magnitude of flow velocity $\mathbf{v}$ and laminar front speed s_L equal to 650. The arsenous-acid/iodate reaction takes place in an aqueous solution and has two advantages over gas reactions: (1) it allows small density changes across the reaction front, (2) it permits large rms values of $\mathbf{v}$ where front normal velocity still depends on local flow and curvature. Both properties are helpful for comparing experimental findings with predictions from the HJ-type front models such as the G-equation. The capillary wave flow is achieved in a thin layer of liquid in a vertically vibrated tray (20 cm on each side). At large amplitude, the flow field becomes random and develops a broad range of spatial and temporal scales. In the experiment, the flow has zero ensemble mean and is isotropic and quasi-two-dimensional. For details of the experimental setup, see [112]. We observe the anisotropic and multiscale features of the front. Image used with the permission of Dr. Paul Ronney.

The G-equation ignores chemical kinetics, front width, and diffusion. The first-principle-based model is the reaction–diffusion-advection equation (or a system of such equations)

$$u_t + \mathbf{v} \cdot \nabla_x u = D \, \nabla^2 u + \frac{1}{\tau} f(u), \qquad (1.5)$$

where u is the concentration of reactant, D is the diffusion constant, and τ is the time scale of reaction. As we shall show in Chapter 5, for the KPP reaction $f(u) = u(1 - u)$, the large time front speed from compactly supported initial data is given by averaging a KPZ equation with random advection (η replaced by $-\mathbf{v} \cdot \nabla_x h$ in (1.3)). On the other hand, the G-equation and its variants are able to approximate front speeds for ignition-type reactions in combustion when the reaction time τ and front width $O(\sqrt{D})$ are much smaller than the time and length scales in $\mathbf{v}$.

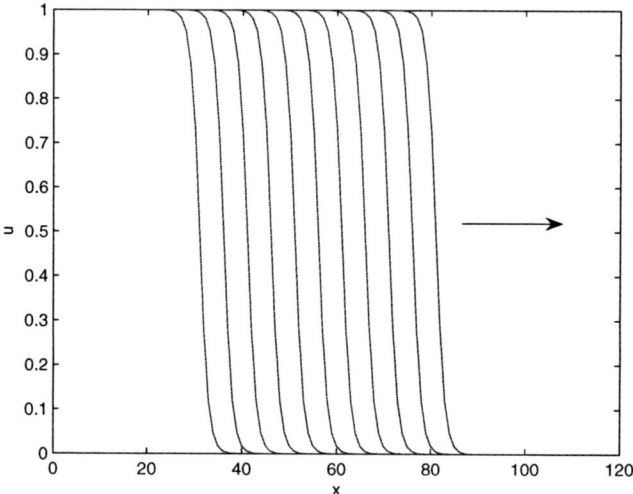

Figure 1.2 Sketch of traveling fronts $U(x - ct)$ moving at constant speed in a homogeneous medium.

Figure 1.1 shows a laser-induced fluorescence light sheet image of an experimental arsenous-acid/iodate autocatalytic front in random capillary wave flows. The reactive front spreads out nonuniformly and takes on a fractal shape. A KPP front spreading is similar, as we shall analyze later. In particular, the front speed variational characterization allows one to estimate and compute the spreading rate in random media. A new theme associated with fronts in heterogeneous media is the understanding of multiple scales and their interaction. We illustrate how to apply homogenization (upscaling) ideas to front problems in periodic and random media. Basic ideas of homogenization theory explained through concrete examples serve as useful guides.

We begin in Chapter 1 with the scalar prototype equations in homogeneous media and explain the basic properties of front solutions. A traveling-front solution in a homogeneous medium is a solution of the form $U(p \cdot x - ct)$, where p is a unit vector, c is a constant speed independent of p, and U is the front profile. Since the front speed and profile are the same in each direction p, it is convenient to consider the traveling front $U(x - ct)$ in the case of one spatial dimension. The scalar conservation laws are partial differential equations (PDEs) of the form

$$u_t + f(u)_x = \nu u_{xx}, \quad (x,t) \in \mathbb{R} \times (0,\infty), \tag{1.6}$$

where f is a smooth nonlinear function, the flux function; and $\nu \geq 0$ is the viscosity parameter. The HJ equations are:

$$u_t + H(u_x) = \nu u_{xx}, \quad x \in \mathbb{R}, \tag{1.7}$$

where H is the Hamiltonian and is nonlinear in u_x (momentum). The R-D equations
are

$$u_t = du_{xx} + f(u), \tag{1.8}$$

where $d > 0$ is the diffusion coefficient and $f(u)$ is the reaction nonlinearity. Figure 1.2 shows a right-moving RD front or a front of a viscous conservation law.

All three equations are well documented in graduate textbooks on PDEs [80, 226]. We shall look at these three nonlinear equations in terms of traveling-front solutions, variational characterizations, front existence and stability, front speed selection, and general solution formulas. Many of these properties carry over to heterogeneous fronts.

1.1 Traveling Fronts of Burgers and Hamilton–Jacobi Equations

For the viscous Burgers equation

$$u_t + uu_x = vu_{xx}, \quad v > 0, \ x \in \mathbb{R}, \tag{1.9}$$

we seek a front solution

$$u(x,t) = U(x - ct) = U(\xi), \quad \xi = x - ct$$

connecting the equilibria of zero and one (see Figure 1.2). Upon substitution, the profile $U(\xi)$ satisfies a second-order ordinary differential equation (ODE), which can be solved exactly under the boundary conditions $U(-\infty) = 1$ and $U(+\infty) = 0$. Details are left as an exercise at the end of the chapter (Section 1.6). The solution is

$$U(\xi) = \frac{1}{1 + \exp\{\xi/2v\}}, \tag{1.10}$$

where x can be shifted by any constant $x_0 \in \mathbb{R}$. The front (1.10) moves to the right at speed $1/2$ without changing its shape and is called a traveling front.

If the initial data $u(x,0) = U(x) + V(x), V(x)$ give a smooth function with enough decay at spatial infinities, a classical result [121] says that solution $u(x,t)$ eventually converges to $U(x - t/2 + x_0)$ uniformly in x for a constant x_0. The constant x_0 depends on the integral (mass) of the initial perturbation $V(x)$. In fact, the Burgers equation conserves the total mass $\int_\mathbb{R} u(x,t) \, dx$. For a bounded and decaying $V(x)$, there is a unique value x_0 such that

$$\int_\mathbb{R} [u(x,0) - U(x + x_0)] \, dx = 0;$$

hence by conservation of mass we have

$$\int_\mathbb{R} \left[u(x,t) - U\left(x - \frac{t}{2} + x_0\right) \right] dx = 0, \quad \forall t > 0.$$

If $V(x)$ is also small, then x_0 is small. Taylor expanding the above equality at $t = 0$ shows that

$$\int_{\mathbb{R}} [u(x,0) - U(x) - U'(x)x_0] \, dx \approx 0,$$

implying

$$x_0 \approx \int_{\mathbb{R}} [u(x,0) - U(x)] \, dx = \int_{\mathbb{R}} V \, dx. \tag{1.11}$$

So for a small perturbation, x_0 is approximately the total mass of the initial perturbation. In the limit $v \downarrow 0$, U_v converges to a step function (shock front) that also travels to the right at speed $1/2$.

The simplest traveling-front solutions of HJ (1.7) are the linear functions

$$u(x,t) = px - H(p)t, \tag{1.12}$$

where p is a nonzero wave number. This is seen by direct substitution in (1.7). They come from spatial integrals of constant solutions of a scalar conservation law. Its level set $u(x,t) = $ constant is a line moving at constant speed. In multiple spatial dimensions, (1.7) extends to $\mathbf{p} \cdot x - H(\mathbf{p})t$.

The level set, $u = $ constant represents a hyperplane moving at speed $H(\mathbf{p})$ in the unit direction $\mathbf{p}$. In one spatial dimension, if $p = 1$, $H(p) = p^2/2$ as in the Burgers equation, the HJ front moves to the right with speed $\frac{1}{2}$. Clearly, there may exist other traveling-front solutions. For example, if $H(u) = u^2/2$, the integral of the Burgers traveling front $-\int_{x-t/2}^{\infty} U(\xi) \, d\xi$ is an HJ traveling front, where U is given by (1.10). For small v, this solution is approximately a moving cone, or the function $x1_{(-\infty,0)}(x)$, with 1_A being the indicator function of the set A. One may view such a traveling front as consisting of two solutions of the form (1.12) with two values of p. The traveling front (1.12) is truly a planar solution, a building block of more complicated solutions.

The Burgers equation (1.9) has a closed-form solution formula, or (1.9) is integrable. By the substitution $u = -2v\varphi_x/\varphi$, we obtain the linear heat equation $\varphi_t = v\varphi_{xx}$ for φ and the well-known Hopf–Cole formula [237] in terms of the heat kernel:

$$u(x,t) = \left(\int_{-\infty}^{\infty} \frac{x - \eta}{t} \exp\{-(2v)^{-1}G(\eta)\} d\eta \right) \left(\int_{-\infty}^{\infty} \exp\{-(2v)^{-1}G(\eta)\} d\eta \right)^{-1}, \tag{1.13}$$

where

$$G(\eta) = G(\eta; x, t) = \int_0^{\eta} u(\eta', 0) \, d\eta' + (2t)^{-1}(x - \eta)^2. \tag{1.14}$$

For a convex Hamiltonian H with superlinear growth, $\lim_{|p| \to \infty} H(p)/p = +\infty$, the Legendre transform

$$L(p) = \sup_{q \in \mathbb{R}} (p \cdot q - H(q)) \tag{1.15}$$

defines the convex Lagrangian L, which also grows superlinearly at large $|p|$. Suppose the initial datum $u(x,0) = g(x)$ is a Lipschitz continuous function such that

$|g(x) - g(y)| \leq L_p |x - y|$ for some constant L_p and all $x, y \in \mathbb{R}$, with the inviscid HJ solution in the limit of $\nu \downarrow 0$ being given by the Hopf formula [80]

$$u(x,t) = \inf_{y \in \mathbb{R}} \left[tL\left(\frac{x-y}{t}\right) + g(y) \right], \quad t > 0, \tag{1.16}$$

which is Lipschitz continuous in $\mathbb{R} \times [0, \infty)$.

The Hopf solution (1.16) is almost everywhere differentiable in (x,t), where it satisfies the HJ equation and its initial condition [80]. Because the Lagrangian L has superlinear growth and g has at most linear growth, the infimum in (1.16) is attained at a finite point $y \in \mathbb{R}$.

The solution formula of the inviscid Burgers equation (or a convex conservation law) can be derived from the Hopf formula. For front initial data $u(x,0)$ with enough decay at plus infinity, define $h(x) = -\int_x^{+\infty} u(x',0)\, dx'$ and $w(x,t) = -\int_x^{+\infty} u(x',t)\, dx'$. Then w solves the HJ equation $w_t + f(w_x) = 0$, $w(x,0) = h(x)$, and the Hopf formula for w gives the solution formula for $u = w_x$. A variant of this representation is called the Lax–Oleinik formula [141, 80], which is also the inviscid limit ($\nu \downarrow 0$) of (1.6); see [80].

1.2 Traveling Fronts of Reaction–Diffusion Equations

The traveling-front solutions to reaction–diffusion equations (RD) (1.8) are special solutions of the form $u = U(x - ct) \equiv U(\xi)$, where c is the front speed and U is the front profile that connects 0 and 1. Substituting this form into (1.8) with $d = 1$, we obtain

$$U_{\xi\xi} + cU_\xi + f(U) = 0, \tag{1.17}$$

with boundary conditions $U(-\infty) = 0$ and $U(\infty) = 1$. Since u is a concentration or a temperature, we also impose the physical condition $U(\xi) \geq 0$.

Note that we could also have an RD front with $U(-\infty) = 1$ and $U(+\infty) = 0$ by simply changing variables $x \to -x$, $c \to -c$ in (1.17), which is invariant. However, for a convex conservation law or the Burgers equation, the front must connect one at minus infinity to zero at plus infinity. Reversing the boundary conditions at infinity will produce the so-called rarefaction (expansion) wave. This is because (1.6) models compressible fluids (air). The front describes air compression physically, and is formed only under a pressure difference (or the pressure is higher at the inlet of a piston than the pressure inside). In contrast, the RD equation (1.8) models a flame if f is a suitable nonlinearity (type 4 or type 5 below, line–circle or line–star in Figure 1.3). A flame moves from a hot ($u = 1$) region to a cold ($u = 0$) region, irrespective of whether the hot region is on the left or on the right side. This is an interesting difference between (1.6) and (1.8). The other is that the front speed of (1.6) is explicit, thanks to the conservation of mass, while that of (1.8) is typically implicit due to the lack of an invariant quantity in the evolution. The similarity of the fronts of (1.6) and (1.8) is that they all carry information on speed and profile.

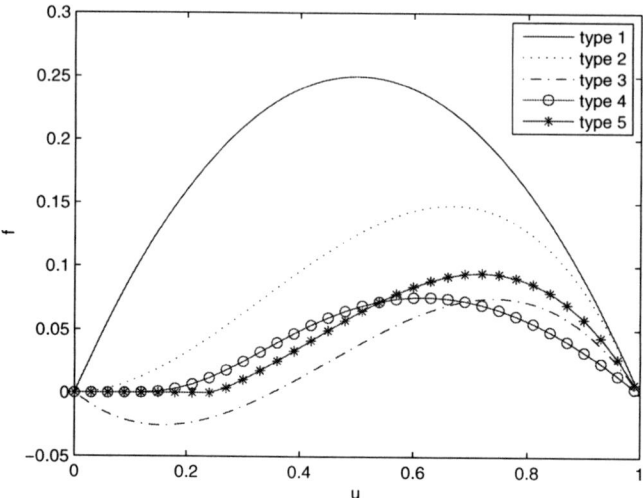

Figure 1.3 Sketch of five types of reactive nonlinearities: type 1 ($u(1-u)$, KPP-Fisher) solid line; type 2 ($u^2(1-u)$, higher-order KPP–Fisher) dotted line; type 3 (bistable) line–dot; type 4 (Arrhenius) line–circle; type 5 (ignition) line–star.

To be specific, for the study of traveling fronts, we will be concerned with the following five types of nonlinearities:

1. $f(u) = u(1-u)$: the Kolmogorov–Petrovsky–Piskunov (KPP) [134] or Fisher nonlinearity [91];
2. $f(u) = u^m(1-u)$, m an integer ≥ 2: the mth-order Fisher nonlinearity (called the Zeldovich nonlinearity if $m = 2$);
3. $f(u) = u(1-u)(u-\mu)$, $\mu \in (0,1)$: the bistable nonlinearity;
4. $f(u) = e^{-(E/u)}(1-u)$, $E > 0$: the Arrhenius combustion nonlinearity or combustion nonlinearity with activation energy E but no ignition temperature cutoff;
5. $f(u) = 0 \, \forall u \in [0,\theta] \cup \{1\}$, $f(u) > 0 \, \forall u \in (\theta,1)$, $f(u)$ Lipschitz continuous: the combustion nonlinearity with ignition temperature θ.

Types 1 and 2 come from chemical kinetics (for example, from autocatalytic reactions [142]), with type 2 being the high-order generalization of type 1. Type 3 comes from biological applications (such as FitzHugh–Nagumo systems [9]) and also more recently from phase field models of solidification in material science [90]. Types 4 and 5 appear in the study of premixed flames in combustion science [28, 238]. Types 1, 2, and 4 are nonnegative and can be recovered as a limit of type 5 as θ tends to zero.

If we look at the graphs of $f(u)$ for the five types in Figure 1.3, we see that they differ near $u = 0$ and behave similarly near $u = 1$. The type-1 nonlinearity has a positive slope at $u = 0$. The type-2 nonlinearity has zero slope (and derivatives up to order $m - 1$ for $m \geq 2$). The type-4 nonlinearity has an exponentially small tail near zero, so all derivatives at zero vanish. Type 5 is identically zero for an interval $u \in [0, \theta]$, i.e., there is no reaction below ignition temperature. Type 3 has a negative slope at $u = 0$, then goes down to a negative minimum, goes up and through an intermediate zero μ, then up to its positive maximum, and finally comes back to its third zero at $u = 1$. Type 3 is the only one that changes sign. Its total area $\int_0^1 f(u)du$ is positive if $\mu \in \left(0, \frac{1}{2}\right)$, zero if $\mu = \frac{1}{2}$, and negative if $\mu > \frac{1}{2}$. If we gradually deform the curve of $f = f(u)$ near $u = 0$ from above the u-axis (type 1) to below (type 3), we can experience all five types of nonlinearities.

The boundary value problem (1.17) can be thought of as a nonlinear eigenvalue problem with eigenvalue c and eigenfunction U. It is convenient to perform a phase plane analysis by writing (1.17) as a first-order system of ODEs,

$$U_\xi = V, \quad V_\xi = -cV - f(U). \tag{1.18}$$

Now we are looking for a trajectory in the phase plane that goes from $(0,0)$ to $(1,0)$. Multiplying both sides of (1.17) by U_ξ and integrating over $\xi \in \mathbb{R}$, we obtain (assuming $\mu \in \left(0, \frac{1}{2}\right)$ in the case of a nonlinearity of type 3)

$$c = -\frac{\int_0^1 f(U)\, dU}{\int_{\mathbb{R}} U_\xi^2\, d\xi} < 0.$$

The linearized system about $U = 0$ is

$$\frac{d}{d\xi}\begin{pmatrix} U \\ V \end{pmatrix} = \begin{pmatrix} 0 & 1 \\ -f'(0) & -c \end{pmatrix}\begin{pmatrix} U \\ V \end{pmatrix}.$$

The eigenvalues of this 2×2 matrix are given by

$$\lambda_{1,2} = \frac{-c \pm \sqrt{c^2 - 4f'(0)}}{2}.$$

In the case of a type-1 nonlinearity, if $c^2 \geq 4f'(0)$ or $c \leq c_1^* \equiv -2\sqrt{f'(0)}$, then $(0,0)$ is an unstable node. In the type-3 case, since $f'(0) < 0$, it follows that $(0,0)$ is a saddle. In either case, a similar linearization at $(1,0)$ shows that $(1,0)$ is always a saddle, thanks to $f'(1) < 0$. Since there is a family of unstable directions going out of an unstable node, and only one direction in or out of a saddle, one can show by isolating the flows in a triangular region in the first quadrant of the U-V plane that there is a connecting trajectory for each $c \leq c_1^*$ for type 1, and a unique connecting trajectory for type 3. Moreover, $U_\xi > 0$ always holds, thanks to the trajectory being in the first quadrant. Since the ODE system (1.18) is autonomous, U is unique only up to a constant translation of ξ.

We shall call the front solution corresponding to $c = c_1^*$ the critical front. The critical front moves at the slowest speed in absolute value, and its asymptotic behavior as $|\xi| \to \infty$ is [8]

$$U(\xi) = \begin{cases} 1 - Ce^{-\beta\xi} + O(e^{-2\beta\xi}), & \xi \to +\infty, \\ (A - B\xi)e^{-c_1^*\xi/2} + O(\xi^2 e^{-c_1^*\xi}), & \xi \to -\infty, \end{cases} \tag{1.19}$$

where A, B, and C are positive constants and $2\beta = -c_1^* - \sqrt{(c_1^*)^2 - 4f'(1)} > 0$. In contrast, the faster fronts with $c < c_1^*$ have exponential decay $O(e^{\text{const} \cdot \xi})$ as $|\xi| \to \infty$ because the two roots $\lambda_{1,2}$ at $(0,0)$ are both simple. The faster fronts decay more slowly than the critical fronts as $U \to 0$.

For the type-3 cubic polynomial, Huxley [212] solved (1.17)–(1.18) exactly:

$$U(\xi) = \frac{1}{1 + e^{-\xi/\sqrt{2}}}, \quad c = \sqrt{2}\left(\mu - \frac{1}{2}\right) \tag{1.20}$$

for $\mu \in (0, \frac{1}{2}]$. If $\mu \in [\frac{1}{2}, 1]$, one simply switches c to $-c$ and ξ to $-\xi$.

For the remaining three types, $f'(0) = 0$, and so $(0,0)$ has an unstable and a neutral direction. More delicate analysis is required. In the type-2 case, one can show that there are a center manifold near $(0,0)$ and a connecting trajectory from the center manifold to the saddle at $(1,0)$ for each $c < c_m^* < 0$. If $c = c_m^*$, the connection goes from the unstable manifold at $(0,0)$ to the saddle. For a type-2 nonlinearity with $m = 2$, the critical front profile approaches zero at the rate $O(e^{-c_2^*\xi})$ as $\xi \to -\infty$, while the profiles of faster fronts approach zero at an algebraic rate $O(\xi^{-1})$ [32]. This is different from the KPP case (1.19). The absolute values of c_m^* decrease with increasing m.

For type-4 and type-5 nonlinearities, a different method using degree theory on finite intervals to construct approximate solutions, then taking their infinite line limit, is much more expedient and robust; see [28, 155]. We will explain this method in detail in the coming sections on fronts in periodic media. The results of [28] and [155] show in particular that in the case of a type-4 nonlinearity, a continuum family of traveling-front solutions exists, one for each $c \le c_0^* < 0$, just as for type 1 and type 2. However, type 5 is different from type 4 in that for each given ignition temperature $\theta > 0$, there is a unique c_θ^* such that a corresponding front profile U exists and is unique up to a constant translation in ξ. We see that type 5 is just like type 3. See [89, 228] for a phase-plane justification of the result.

The degree approach is very effective in proving the existence of traveling waves in multiple dimensions. For example, the existence of traveling fronts in channel domains $\mathbb{R} \times \Omega$, Ω is a bounded domain with Lipschitz continuous boundary in $\mathbb{R}^n$, $n \ge 1$. The front moves along the channel and has the form $U(x - ct, y)$, $y \in \Omega$, due to the y-dependent coefficient in the equation (equation (2.19)); see [27, 31] on the existence theory for all five types of nonlinearities. The other related method for existence of traveling fronts is the Conley index theory [226]; see [99] for its application to the existence of fronts in a channel domain.

The next question is the asymptotic stability of traveling fronts in large time. The stability means that if the initial data are prescribed in the form $u_0(x) = U_c(x) + u'(x)$, where $U_c(x)$ is a front profile corresponding to the speed c and $u'(x)$ is a smooth and spatially decaying perturbation, then $u(x,t)$ converges to $U_c(x + ct + \xi_0)$ in a proper function space as $t \to \infty$ for some constant ξ_0. The reason we have a constant translation in the definition can be seen as follows. Due to the spatial translational invariance of the original equation, we have a family of traveling fronts $U_c(x - ct + x_0)$ for each allowable wave speed c. Let us take $u'(x) = U_c(x + x_0) - U_c(x)$, which is a perturbation with spatial decay. Now for initial data $u(x,0) = U_c(x) + u'(x) = U_c(x + x_0)$, the solution for later time is just $U_c(x + ct + x_0)$, which does not converge to $U_c(x + ct)$ unless $x_0 = 0$. In the case of a continuum of speeds, we can also take $u'(x) = U_{c'}(x) - U_c(x)$, $c' \neq c$, and the later-time solution is $U_{c'}(x - c't)$, again not converging to $U_c(x - ct)$ as $t \to \infty$. Even the wave speed is different. The convergence to a translated front is similar to Burgers' equation except that the translated amount is implicit, while for (1.6) it is the total mass of the initial (small) perturbation (1.11).

These simple examples show that it is a subtle problem to establish asymptotic stability, especially in the case of multiple speeds. Much turns out to depend on the rate of decay of the initial perturbations as $|\xi| \to \infty$. Intuitively, the tiny amount of perturbation in the far field takes a long time to crawl into a front from its tails; however, its effect is crucial, since asymptotic stability concerns large-time behavior. Asymptotic stability for noncritical fronts based on the spectral theory of linearized operators can be found in [212]. For asymptotic stability of critical fronts with minimal speeds, see [131] on KPP nonlinearity, [42] on Ginzburg–Landau (GL) nonlinearity $f(u) = u(1 - u^2)$, [98] on both GL and KPP, and [75] on more general parabolic equations. For the global asymptotic stability and critical front selection based on analysis using maximum principles, see [8, 9, 89, 90, 124], the original paper [134] with the initial data being the indicator function of the negative line, and for a probabilistic analysis of the KPP equation, see [40, 95, 160].

Large-time asymptotic stability is also proved for traveling fronts in channel domains; see [205, 15], where in particular the critical front stability problem of KPP for type-2 and type-4 nonlinearities is resolved.

1.3 Variational Principles of Front Speeds

Since the speeds of the RD fronts are in general unknown in closed form, the variational characterization is an invaluable way of estimating them. For general continuously differentiable nonlinearity $f(u)$ such that

$$f(0) = f(1) = 0, \quad f(u) > 0, \ u \in (0,1), \quad f'(0) > 0, \quad f'(1) < 0, \qquad (1.21)$$

the min-max variational principle for the minimum speed was first established [107]:

$$|c_*| = \inf_{\rho} \ \sup_{u \in (0,1)} \left\{ \rho'(u) + \frac{f(u)}{\rho(u)} \right\}, \qquad (1.22)$$

where ρ is any continuously differentiable function on $[0,1]$ such that

$$\rho(u) > 0, \quad u \in (0,1), \quad \rho(0) = 0, \quad \rho'(0) > 0. \qquad (1.23)$$

The formula (1.22) is based on the phase-plane construction of the fronts. Under (1.21), for each allowable c, there is a connection from an unstable node to a saddle, and the front profile is strictly monotone. Let $u = u(x - ct) = u(\xi)$ connect $u = 1$ and $u = 0$ from left to right, so that $c \geq c_* > 0$. Then $u_{\xi\xi} + cu_\xi + f(u) = 0$, $u(-\infty) = 1$ and $u(\infty) = 0$. Now regard u_ξ as a function of u by defining $p = p(u) = -u_\xi > 0$ at $u = u(\xi)$. The function $p(u)$ is a solution of

$$p(u)p'(u) - cp(u) + f(u) = 0, \tag{1.24}$$

with $p(0) = 0$, $p(1) = 0$, and $p(u) > 0$ on $(0, 1)$. The expression inside the supremum of (1.22) is just what we find from (1.24) on writing c in terms of u, p, and $p'(u)$. If $p(u)$ is not a solution of (1.24), then $(u, p(u))$ can represent a curve connecting the node and the saddle in the phase plane, but with the flow field at the curve pointing toward the solution curve of (1.24). This geometric information translates into the inequality that c is no less than the supremum in (1.22) for some p satisfying (1.23). It follows that any allowable c, in particular c_*, is no less than the min-max of (1.22). The equality is attained by $p = p_*(u)$ corresponding to the speed c_*.

It follows from (1.22) that

$$2\sqrt{f'(0)} \leq |c_*| \leq 2\sqrt{L}, \quad L = \sup_{u \in (0,1)} \frac{f(u)}{u}, \tag{1.25}$$

which gives the well-known KPP minimal speed $2\sqrt{f'(0)}$ if L is achieved at $u = 0$. To see (1.25), we take $\rho(u) = au$, $a > 0$. Then $|c_*| \leq a + L/a$. Minimizing over a establishes the upper bound. The lower bound is easily deduced by restricting the supremum to those functions u in a small neighborhood of zero. The formula (1.22) was used further in [107] to find the exact minimal speed for $f(u) = u(1 - u)(1 + vu)$: $|c_*| = 2$ if $-1 \leq v \leq 2$, $|c^*| = (v + 2)/\sqrt{2v}$ if $v \geq 2$.

Recently, a general variational speed formula was found [17] for any f such that $f(0) = f(1) = 0$. Let f be any of the five types of nonlinearity, and assume that a monotone front exists. Then the minimum (or unique) speed c_* is given by

$$c_*^2 = \sup \left(2 \frac{\int_0^1 fg\,du}{\int_0^1 (-g^2/g')\,du} \right), \tag{1.26}$$

where the supremum is over all positive decreasing functions $g \in (0, 1)$ for which the integrals exist. Moreover, the maximizer exists if $|c_*| > 2\sqrt{f'(0)}$. The formula (1.26) appears to be the first variational result in such generality. The formula holds for f changing signs in $(0, 1)$. The Huxley formula (1.20) is recovered by putting $g(u) = ((1 - u)/u)^{1-2\mu}$. A similar variational formula for f in (1.21) without the constraint $f'(1) < 0$ is established in [16].

The proof of (1.26) is elementary, and it uses (1.24) again. Let $g = g(u)$ be any positive function on $(0, 1)$ such that $h = -g'(u) > 0$. Multiplying (1.24) by $g(u)$ and integrating over $u \in [0, 1]$, we have after integration by parts the equality

$$\int_0^1 fg\,du = c\int_0^1 pg\,du - \int_0^1 \frac{1}{2}hp^2\,du. \qquad (1.27)$$

For positive c, g, and h, the function $\varphi(p) = cpg - \frac{1}{2}hp^2$ has its maximum at $p = cg/h$, and so $\varphi(p) \le c^2g^2/2h$. It follows that

$$c^2 \ge 2\frac{\int_0^1 fg\,du}{\int_0^1 (g^2/h)\,du} \equiv I(g), \qquad (1.28)$$

which implies (setting $c = c_*$ if c is nonunique) that c_*^2 is no less than the supremum of (1.26). Next, we show that the equality holds for a function $\hat{g}$. Notice that the condition $p = cg/h$ is solvable in g and gives an expression for the maximizer $\hat{g}$,

$$\hat{g} = \exp\left\{-\int_{u_0}^u cp^{-1}\,du\right\}, \qquad (1.29)$$

with $u_0 \in (0,1)$. Clearly, $\hat{g}$ is positive and decreasing, with $\hat{g}(1) = 0$, since $p \sim O((1-u))$ for $u \sim 1$. At $u = 0$, however, $\hat{g}$ diverges, since the exponent goes to $+\infty$. A natural choice for $\hat{g}$ now is $p = p_*(u)$ if we verify that the two integrals are finite in $I(\hat{g})$. For nonlinearities of types 2, 4, and 5, p_* approaches zero exponentially and

$$p_* \sim \frac{c + \sqrt{c^2 - 4f'(0)}}{2}u \equiv mu.$$

Thus, near $u = 0$, $\hat{g} \sim u^{-c/m}$, and $f\hat{g}$ and $\hat{g}^2/\hat{h}$ diverge at most like $u^{1-c/m}$. The integrals of $I(\hat{g})$ are finite if $m/c > \frac{1}{2}$. This condition holds if $f'(0) \le 0$, which is indeed true for types 2, 4, and 5, and also for f in (1.21) if $c_*^2 > 4f'(0)$. If $c_*^2 = 4f'(0)$, which is the case for type 1, the maximizer does not exist. However, choosing the test function $g(u) = a(2-a)u^{a-2}$ with $a \in (0,1)$, we calculate $\int_0^1 (g^2/h)\,du = 1$. Integration by parts twice shows that as $a \to 0$, $I(g) = 2(2-a)a\int_0^1 fu^{a-2}\,du \to 4f'(0)$. The proof is complete.

1.4 Random Variables and Stochastic Processes

In this section, we give a brief introduction to random variables and stochastic processes as a preparation for later chapters. We shall follow [41, 72, 206] for basic definitions and concepts in probability and stochastic processes below, and refer to [55, 206] for more examples and applications. First, a *probability space* is a triple $(\Omega, \mathscr{F}, \mathscr{P})$, where (1) Ω is a set of outcomes (sample space), a subset of which is called an event; $\varnothing$ is the impossible event; (2) $\mathscr{F}$ is a collection of events including Ω and $\varnothing$ with the property that $\mathscr{F}$ is closed under complementation, countable intersections, and unions; (3) P is a probability function that assigns probability to events in $\mathscr{F}$. A *probability* is a nonnegative set function defined on $\mathscr{F}$ with values in $[0,1]$ such that (a) (normalization) $P(\Omega) = 1$, $P(\varnothing) = 0$; (b) (additivity) for any finite or countable disjoint collection B_k of sets in $\mathscr{F}$, $P(\cup_k B_k) = \sum_k P(B_k)$.

As a finite sample space example, consider throwing a die. There are six possible outcomes, denoted by ω_i, $i = 1,\ldots,6$. The set of all outcomes is the sample space $\Omega = \{\omega_1,\ldots,\omega_6\}$. A subset of Ω, an event, is $A = \{\omega_2,\omega_4,\omega_6\}$. Suppose we did N die experiments, and event A happened N_a times. The *probability* of event A is $P(A) = \lim_{N\to\infty} N_a/N$. For a fair die, $P(A) = \frac{1}{2}$. An infinite sample space example is $\Omega = (0,1)$, $\mathscr{F}$ is the collection of all open subsets and their countable unions in $\mathbb{R}$, P is the Lebesque measure with $P((a,b]) = b - a$ for all $a < b$. See [72, 55] for more examples.

A real-valued function X defined on Ω is a *random variable*. For example, an indicator function of a set $A \in \mathscr{F}$ is a random variable (r.v.). The *distribution function* of an r.v. is $F(x) = P(X \leq x)$, which is nondecreasing and right continuous in x, and $F(+\infty) = 1$, $F(-\infty) = 0$. If $F(x)$ is absolutely continuous, the *density function* is $f(x) = F'(x)$. The *expectation* of X is

$$E[X] \equiv \mu = \int_{-\infty}^{+\infty} x f(x)\,dx,$$

and the *variance* of X is

$$\mathrm{Var}[X] = E[(X - \mu)^2] \equiv \sigma^2,$$

where σ is the *standard deviation*.

A unit Gaussian r.v. (or a standard normal r.v.) is described by the density function

$$f(x) = (2\pi)^{-1} \exp\{-x^2/2\},$$

with $\mu = 0$, $\sigma = 1$. The density function of a uniformly distributed r.v. on $(0,1)$ is the indicator function of the interval $(0,1)$, with $\mu = 1/2$, $\sigma^2 = 1/12$.

Random variables $X_1, X_2, \ldots, X_n$ are independent if for any $a_i \in \mathbb{R}$ $(i = 1,2,\ldots,n)$,

$$P(\cap_{i=1}^n \{\omega : X_i(\omega) \leq a_i\}) = \prod_{i=1}^n P(\{\omega : X_i(\omega) \leq a_i\}).$$

Likewise, events $A_1,\ldots,A_n$ are independent if

$$P(\cap_{i=1}^n A_i) = \prod_{i=1}^n P(A_i).$$

The countably many random variables $X_1, X_2, \ldots$, are independent if for every $n \geq 2$, the random variables $X_1, X_2, \ldots, X_n$ are independent.

Concerning the limit behavior of a sequence of random variables $X_1, X_2, \ldots, X_n, \ldots$, a few commonly used modes of convergence are as follows:

- (cp1) Convergence with probability one (almost sure convergence) if there exists an r.v. X such that

$$P\left(\{\omega \in \Omega : \lim_{n \to \infty} |X_n(\omega) - X(\omega)| = 0\}\right) = 1. \tag{1.30}$$

- (msc) Mean-square convergence if $E(X_i^2) \leq C$ for a constant C, and

$$\lim_{n \to \infty} E(|X_n - X|^2) = 0. \tag{1.31}$$

- (cp) Convergence in probability if

$$\lim_{n \to \infty} P(\{\omega \in \Omega : |X_n(\omega) - X(\omega)| \geq \varepsilon\}) = 0, \ \forall \varepsilon > 0. \tag{1.32}$$

- (cl) Convergence in law if there exists an r.v. such that

$$\lim_{n \to \infty} F_{X_n}(x) = F_X(x) \tag{1.33}$$

at all continuous points of F_X (the distribution function of X). We shall denote the limit in this sense of convergence or equivalence in the sense of distributions as $\overset{\text{law}}{=}$ or $\overset{\text{d}}{=}$.

General inference relations are cp1 (msc) $\implies$ cp $\implies$ cl.

The fundamental laws of probability on a sequence of independent identically distributed (iid) random variables X_i can be stated as follows. Suppose $E|X_i| < \infty$, and set $\mu = E(X_i)$, $\sigma^2 = Var(X_i) \in (0, \infty)$. The strong law of large numbers refers to the convergence of

$$\frac{S_n}{n} = \frac{\sum_{i=1}^{n} X_i}{n} \to \mu \tag{1.34}$$

in the sense of (wp1) and (msc). The weak law of large numbers refers to (1.34) in the sense of (cp). The central limit theorem (CLT)

$$\frac{S_n - n\mu}{\sigma \sqrt{n}} \to N(0, 1) \tag{1.35}$$

holds in law (cl), where $N(0, 1)$ is the unit Gaussian.

A sequence of random variables $X_1, X_2, \ldots, X_n, \ldots$ occurring at discrete times $t_1 < t_2 < \cdots < t_n$ is called a *discrete stochastic process*, with joint distribution $F_{X_{i_1}, X_{i_2}, \ldots, X_{i_j}}$ as its probability law. The process is called Gaussian if all joint distributions are Gaussian.

A *continuous stochastic process* $X(t) = X(t, \omega)$, $t \in I$ an interval of $\mathbb{R}$, over the probability space (Ω, A, P), is a function of two variables $X : I \times \Omega \to \mathbb{R}$, where X is an r.v. for each t; for each ω, we have a sample path (a realization) or trajectory of the process.

A few basic quantities are $\mu(t) = E(X(t, \omega))$, $\sigma^2(t) = Var(X(t, \omega))$, and the covariance function

$$C(s, t) = E((X(s, \omega) - \mu(s))(X(t, \omega) - \mu(t))),$$

for $s \neq t$. The covariance measures how correlated the two random variables are at two points of "time" s and t.

The standard *Wiener Process (Brownian motion)* is a Gaussian process $W(t, \omega)$, $t \geq 0$, with independent increment, and

$$W(0) = 0 \text{ w.p. } 1, \quad E(W(t)) = 0, \quad \text{Var}(W(t) - W(s)) = t - s, \quad (1.36)$$

for all $s \in [0,t]$. It follows that for any $t_0 < t_1 < \cdots < t_n$, the random variables $W(t_k) - W(t_{k-1})$ are independent normally distributed with mean zero and $E[(W(t_k) - W(t_{k-1}))^2] = t_k - t_{k-1}$. In particular, the covariance function $C(s,t)$ is equal to $\min(s,t)$.

A few useful properties of Wiener process are (1) (regularity) almost surely, the sample path $W(t, \omega)$ is Hölder continuous with exponent less than $\frac{1}{2}$, hence nowhere differentiable; (2) (scaling) the process $\tilde{W}(t) = tW(1/t)$ if $t > 0$; $\tilde{W}(0) = 0$ is also a Wiener process; (3) (large-time behavior) $W(t, \omega)$ oscillates and grows sublinearly at large t; its upper and lower envelopes obey the law of the iterated logarithm, that is, almost surely in ω,

$$\limsup_{t \to \infty} \frac{W(t)}{\sqrt{2t \log(\log t)}} = 1, \quad (1.37)$$

$$\liminf_{t \to \infty} \frac{W(t)}{\sqrt{2t \log(\log t)}} = -1. \quad (1.38)$$

The sublinear growth of $W(t)$ is a consequence of its independent increments and the law of large numbers. If t is a positive integer, write

$$W(t) = (W(t) - W(t-1)) + (W(t-1) - W(t-2)) + \cdots$$
$$+ (W(2) - W(1)) + (W(1) - W(0))$$

as a sum of iid random variables with mean zero and variance one, and so $W(t)/t \to 0$ almost surely. For general t, the same limit follows as $|W(t) - W([t])| \leq C_\omega$ by continuity and independent increment properties of W, where $[t]$ is the integral part of t.

A stochastic process is stationary if all joint distributions are translation-invariant. A Gaussian process is stationary if only its covariance function is translation-invariant.

The Wiener process is Gaussian but not stationary. A well-known stationary Gaussian process is the *Ornstein–Uhlenbeck process* (O-U), which is defined as a Gaussian process with $X(0)$ a unit Gaussian, $E(X(t)) = 0$, and the covariance function $E(X_s X_t) = e^{-\gamma|t-s|}$ for $s,t \in \mathbb{R}$, for some constant $\gamma > 0$. Another example is Gaussian white noise, formally $W'(t)$, the derivative of a Wiener process. One may construct it by passing to the limit on an approximate Gaussian process $X_h(t) = (W(t+h) - W(t))/h$, for $h > 0$ small. The process X^h has covariance

$$C_h(s,t) = \frac{1}{h} \max\left(0, 1 - \frac{|t-s|}{h}\right),\tag{1.39}$$

whose Fourier transform (called spectral density) is

$$\frac{\sin^2(2\pi\lambda h)}{(\pi\lambda h)^2} \equiv F_h'(\lambda).\tag{1.40}$$

In the limit $h \to 0$, C_h converges to a delta function, and F_h' converges to a constant (flat spectrum or white color). Accordingly, the process X_h converges in some weak sense to the white noise process.

Wiener and O-U processes all belong to a class of *Markov processes* called *diffusion processes*. Suppose that the multipoint joint distribution function of $X(t)$ has density $p(t_1,x_1;t_2,x_2;\dots;t_k,x_k)$, and define the conditional probability

$$P(X(t_{n+1}) \in B|X(t_i) = x_i, i = 1:n) = \frac{\int_B p(t_1,x_1,\dots,t_n,x_n;t_{n+1},y)\,dy}{\int p(t_1,x_1,\dots,t_n,x_n;t_{n+1},y)\,dy}\tag{1.41}$$

for B any open set of $\mathbb{R}$. The process is *Markov* if

$$P(X(t_{n+1}) \in B|X(t_i) = x_i, i = 1:n) = P(X(t_{n+1}) \in B|X(t_n) = x_n),$$

and the transition probability is

$$P(s,x;t,B) = \int_B p(s,x;t,y)\,dy,$$

where p is the transition density. A Markov process with transition density is called a *diffusion process* if the following limits exist for any $\varepsilon > 0$:

$$\lim_{t\to s^+} \frac{1}{t-s} \int_{|y-x|>\varepsilon} p(s,x;t,y)\,dy = 0,$$

$$\lim_{t\to s^+} \frac{1}{t-s} \int_{|y-x|\le\varepsilon} (y-x)p(s,x;t,y),dy = a(s,x),$$

$$\lim_{t\to s^+} \frac{1}{t-s} \int_{|y-x|\le\varepsilon} (y-x)^2 p(s,x;t,y)dy = b^2(s,x).$$

Alternatively, we may write

$$a(s,x) = \lim_{t\to s^+} \frac{1}{t-s} E(X(t) - X(s)|X(s) = x),$$

$$b^2(s,x) = \lim_{t\to s^+} \frac{1}{t-s} E((X(t) - X(s))^2|X(s) = x).\tag{1.42}$$

The function a is called the *drift coefficient*, and b is the *diffusion coefficient*. Drift and diffusion coefficients indicate the rates of infinitesimal motion of the process over slow (diffusion) and fast (drift) time scales. Using the definitions of Wiener and O-U processes, one calculates that $(a,b) = (0,1)$ for a Wiener process, and $(a,b) = (-\gamma x, 2\gamma)$ for O-U. Over a small time interval $[s,t]$, using drift–diffusion information, we see that O-U is related to W as

$$X(t) - X(s) = -\gamma X(s)(t - s) + \sqrt{2\gamma}(W(t) - W(s)),$$

or in differential form,

$$dX = -\gamma X \, dt + \sqrt{2\gamma} \, dW, \qquad (1.43)$$

a stochastic differential equation (SDE), also known as the Langevin equation. The term $-\gamma X dt$ introduces damping on Brownian motion. For more discussion of SDE and Brownian motion, see [125].

If the initial distribution of O-U at $t = 0$ is a Gaussian (normal) r.v. with mean zero and variance $\rho > 0$, or $\mathcal{N}(0,\rho)$, then X is a stationary Gaussian process with covariance $C(t,s) = \rho \, e^{-\gamma|t-s|}$.

1.5 Noisy Burgers Fronts and the Central Limit Theorem

As an application of stochastic processes to fronts, we study the effects of initial white noise perturbations of Burgers fronts. This is a step beyond the deterministic localized perturbations. In this class of problems, the noise enters initially and the governing equation (1.6)–(1.8) remain the same as in classical front stability analysis.

Consider the Burgers equation (1.9) with initial data

$$u(x,0) = \frac{1}{1 + \exp\{x/(2v)\}} + V(x), \qquad (1.44)$$

where $V(x)$ is either white noise or a stationary Gaussian process with enough decay of correlations.

Now $V(x)$, being a stationary random process, has no decay at infinities. It turns out that at time t, the truncated mass of $V(x)$, or the integral of $V(x)$, over the interval $\left[-\frac{1}{2}t, \frac{1}{2}t\right]$ plays the role of the whole line integral (1.11) and causes the deviation of front location from the mean position $\frac{1}{2}t$. The "one-half" comes from the unperturbed front speed, and the interval $\left[-\frac{1}{2}t, \frac{1}{2}t\right]$ resembles the domain of dependence for the linear wave equation $u_{tt} - 4^{-1}u_{xx} = 0$. The picture behind this is that the perturbation gets sucked into the front from left and right at speed one-half. Let us calculate formally the front deviation for white noise (formally W_x) as in (1.11):

$$x_0 = x_0(t, \omega) \sim \int_{-t/2}^{t/2} W_x(x) \, dx = W(t/2) - W(-t/2)$$

$$\overset{\text{law}}{=} W_t \overset{\text{law}}{=} \sqrt{t} W_1, \qquad (1.45)$$

that is, $\sqrt{t}$ times the unit Gaussian. Thus the front location is

$$X = X(t,\omega) = \frac{t}{2} + x_0(t,\omega) \stackrel{\text{law}}{=} \frac{t}{2} + \sqrt{t}W(1). \tag{1.46}$$

Figure 1.4 (top) illustrates a random front moving according to the law $X(t,\omega) = ct + \tilde{W}(t,\omega)$, uniformly sampled in time (100 time slices) with the corresponding suitably scaled velocity (bottom) at the sampled times. The noise term $\tilde{W}$ is a numerical approximation of the Wiener process W. The constant c is positive and nonrandom. Compared with the uniform speed motion in Figure 1.2, the front speed in Figure 1.4 is highly oscillatory and random-looking.

The above heuristics are made precise in [233]:

Theorem 1.1. *Let $u(x,t)$ be the solution to the initial value problem of the Burgers equation (1.9) and (1.44). Let f be an increasing function of t. Then we have the following:*

1. *(Front Probing) If $\frac{f(t)-t/2}{\sqrt{t}} \to c \in \mathbb{R}$, then $u(f(t),t)$ converges in distribution to a random variable equal to zero with probability $\mathcal{N}(c)$ and equal to one with probability $1 - \mathcal{N}(c)$, where*

$$\mathcal{N}(c) = \frac{1}{\sqrt{2\pi}} \int_{-\infty}^{c} \exp\{-y^2/2\}\,dy$$

is the unit Gaussian distribution function.
Given any positive number $\varepsilon \in (0,1)$, let us define the left and right endpoints of the interval containing the front as

$$z_-(t) = \min\{x : u(t,x) = 1 - \varepsilon\}, \quad z_+(t) = \max\{x : u(t,x) = \varepsilon\},$$

and so the front width is $\{z_+(t) - z_-(t)\}$. Then:
2. *(Front Width) There exists a constant $t_0 > 0$ such that the random variables $\{z_+(t) - z_-(t)\}$ are tight for $t \geq t_0$; i.e., for any $\delta > 0$, there exists an M such that $\mathrm{Prob}(z_+(t) - z_-(t) > M) < \delta$ for all $t > t_0$.*
3. *(Front Motion) As $s \to \infty$, there is a constant σ depending only on $V(x)$ ($\sigma = 1$ for white noise) such that (the same is true for z_-)*

$$\frac{z_+(t) - t/2}{\sigma\sqrt{t}} \stackrel{\text{law}}{\to} W(1).$$

Part 1 is a slightly weaker version of 3, and both substantiate the formal calculation. Part 2 says that the noise does not spread the front width for large time, so nonlinearity dominates over the randomness and preserves the coherent structure. The proof uses the Hopf–Cole formula and a Laplace method for stochastic integrals [233].

Stability of other wave solutions in Burgers and convex conservation laws can be found in [85, 235]. If one performs the hyperbolic scaling change of variables $x \to x/\varepsilon$, $t \to t/\varepsilon$, ε small, the inviscid convex conservation law (1.6) is invariant. Suppose the unperturbed initial datum is the indicator function $1_{\mathbb{R}^\pm}(x)$, the unit step function supported on the half-line $\mathbb{R}^\pm$. Such data lead to a front (minus sign) or

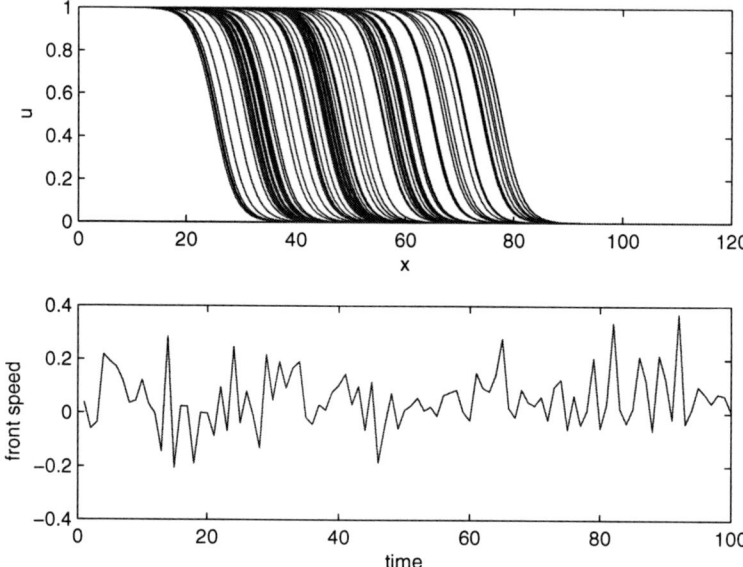

Figure 1.4 Sketch of one hundred uniformly sampled time slices of random fronts $U(x - X(t, \omega))$, $X(t, \omega) = ct + \tilde{W}(t, \omega)$ (top); with the corresponding suitably scaled front speed (bottom). The noise $\tilde{W}$ is a numerical approximation of the Wiener process W. The front profile U is nonrandom and invariant in time, and c is a positive nonrandom constant.

a rarefaction wave (plus sign) corresponding to compression and decompression of gas in a piston. Now we perturb the initial data with white noise, so $u(x, 0) = 1_{\mathbb{R}^\pm}(x) + W_x(x)$. After the scaling change of variables, the initial datum is $u^\varepsilon(x, 0) = 1_{\mathbb{R}^\pm}(x) + W_x(x/\varepsilon)$. Write the scaled solution $u^\varepsilon(x, t) = v_x^\varepsilon$, where v^ε is the Hopf solution to the Hamilton–Jacobi equation:

$$v_t^\varepsilon + f(v_x^\varepsilon) = 0, \quad v^\varepsilon(x, 0) = x \, 1_{\mathbb{R}^\pm}(x) + \varepsilon W(x/\varepsilon). \tag{1.47}$$

Because of the nearly square-root growth of W, the scaled perturbation $\varepsilon W(x/\varepsilon)$ goes to zero for x on any finite interval. In fact, using properties of a Wiener process and the Hopf formula, it can be shown [235] that with probability one, u^ε converges in the sense of distributions to the unperturbed solution of (1.6), which is either the unperturbed shock (minus sign) or a rarefaction wave (plus sign). Hence both waves are stable in the sense of the hyperbolic limit.

The result can be extended for colored noise (stationary Gaussian processes with decaying correlations) [235]. However, the slower diffusive motion of the front is not seen in this limit. Likewise, more detailed stability for a rarefaction wave requires a large-time asymptotic analysis of u; see [85] for the Burgers case. For a mathematical analysis of random Burgers and KPZ equations in connection with turbulence, see [73, 239].

1.6 Exercises

1. Find the Burgers traveling-front formula (1.10) by letting $u = U(x - ct)$ in the Burgers equation (1.9), deriving a second-order ordinary differential equation for U and solving it under the boundary condition $U(-\infty) = 1$, $U(+\infty) = 0$.

2. Derive the Hopf–Cole solution formula (1.13)–(1.14) by writing down a solution to the heat equation $\varphi_t = \nu \varphi_{xx}$, then setting $u = -2\nu \varphi_x / \varphi$. Find the correspondence between the initial data of the Burgers equation and the heat equation.

3. Verify Huxley's traveling-front formula (1.20) for the bistable reaction–diffusion equation (1.8) with $d = 1$ and $f(u) = u(1 - u)(u - \mu)$, $\mu \in \left(0, \frac{1}{2}\right]$. Then generalize the formula to the case $d > 0$ and study how the diffusion constant d influences the solution.

4. Use the conditional expectation formula (1.42) to show that the drift and diffusion coefficients of the Wiener process W are equal to $(0, 1)$. Likewise, for the O-U process $X(t)$, derive its drift and diffusion coefficients using the fact that the increment $X(t + s) - e^{-\gamma s} X(t)$ is independent of the past or events in $\mathscr{F}(X(\tau), \tau \leq t)$.

Chapter 2
Fronts in Periodic Media

Fronts or interfaces in periodic media are deterministic problems in between homogeneous media and random media. Much can be learned on how front solutions
transition from monoscale simple solutions in Chapter 1 to multiple-scale solutions.
Periodic homogenization and PDE techniques based on maximum principles are es-
sential tools for constructing front solutions and analyzing their asymptotics. We
shall observe the close relationship between Hamilton–Jacobi (HJ) and reaction–
diffusion (RD) equations, and present the variational principles of front speeds.

2.1 Periodic Media and Homogenization

Multiscale problems are common in applications such as finding the effective con-
ductivity of a composite material or the effective permeability for flows in porous
media, where one has at least two scales, the large scale of the sample and the small
scale of the embedded inclusions or pores. These two scales normally differ signifi-
cantly and render the full resolution of the problem difficult. Therefore, it is of great
theoretical and practical interest to find out how to upscale the collective effect of the
small scale into the large scale and simplify the problem. When the small scale pos-
sesses a periodic structure, the upscale problem has a well-developed theory called
homogenization. See [18] for a systematic account of the foundational works.

We give here an example of homogenization and use formal asymptotic analysis
to illustrate the ideas. Consider a two-point boundary value problem of a second-
order ODE with rapidly oscillating periodic coefficients,

$$(a(\varepsilon^{-1}x)u_x^\varepsilon)_x = f(x), \quad x \in [0,1], \tag{2.1}$$

with boundary condition $u^\varepsilon(0) = u^\varepsilon(1) = 0$. Here a is a positive smooth function
with period 1 in $y \equiv \varepsilon^{-1}x$, and $f(x)$ is a bounded continuous function in x. We are
going to examine the limit of u^ε as $\varepsilon \to 0$, where the large-scale x and small-scale
$\varepsilon^{-1}x$ are separated. Since there are two separate scales in the problem, it is natural

J. Xin, *An Introduction to Fronts in Random Media*, Surveys and Tutorials in the Applied
Mathematical Sciences 5, DOI: 10.1007/978-0-387-87683-2_2,
© Springer Science + Business Media, LLC 2009

to search for a two-scale expansion of the solution in the form

$$u^{\varepsilon} \sim u_0(\varepsilon^{-1}x) + \varepsilon u_1(x, \varepsilon^{-1}x) + \varepsilon^2 u_2(x, \varepsilon^{-1}x) + \cdots, \tag{2.2}$$

where the $y = \varepsilon^{-1}x$ dependence has period 1 also. Substituting the ansatz (2.2) into (2.1), and regarding x and y as independent variables, we have (noting that the x derivative is replaced by the operator $\partial_x + \varepsilon^{-1}\partial_y$)

$$(\partial_x + \varepsilon^{-1}\partial_y)(a(y)(\partial_x + \varepsilon^{-1}\partial_y)(u_0 + \varepsilon u_1 + \varepsilon^2 u_2 + \cdots)) = f. \tag{2.3}$$

At the highest order $O(\varepsilon^{-2})$, we have

$$\partial_y(a(y)\partial_y u_0) = 0, \tag{2.4}$$

which has only a y-independent periodic solution. Thus $u_0 = u_0(x)$. At the next-highest order $O(\varepsilon^{-1})$, we have

$$\partial_y(a(y)(\partial_x u_0 + \partial_y u_1)) = 0, \tag{2.5}$$

which implies

$$a(y)(\partial_x u_0 + \partial_y u_1) = c(x) \tag{2.6}$$

for some function $c(x)$. Dividing (2.6) by a and integrating the resulting equation over $y \in [0,1]$ yields

$$\frac{d}{dx}u_0 = c(x)\langle a^{-1} \rangle, \tag{2.7}$$

where $\langle \cdot \rangle$ denotes the integral or average over $y \in [0,1]$. At the next order $O(1)$, we have

$$\partial_x(a(y)(\partial_x u_0 + \partial_y u_1)) + \partial_y(a(y)(\partial_x u_1 + \partial_y u_2)) = f. \tag{2.8}$$

Averaging (2.8) over $y \in [0,1]$ gives

$$\partial_x \langle a(y)(\partial_x u_0 + \partial_y u_1)) \rangle = f,$$

which in view of (2.6) is just $dc/dx = f$. This then becomes, when we insert (2.7),

$$\frac{d}{dx}\left(a^* \frac{d}{dx}u_0\right) = f, \tag{2.9}$$

where $a^* = \langle a^{-1} \rangle^{-1}$ is the harmonic mean of a. Equation (2.9) is the homogenized equation and is the same type of equation from which we started; however, its coefficient has been changed to the harmonic mean of the original one in the rapidly oscillating variable $y = \varepsilon^{-1}x$. Now we have only to solve the large-scale equation (2.9) subject to the same boundary condition, and the small-scale effect has been built in already.

Rigorous justifications of the above formal asymptotics in any number of dimensions are presented in [18] using the energy method and in [78] using the weak convergence method; see [190] for the first homogenization result in random media (*a* is a bounded positive random matrix). Equation (2.5) is posed on the periodic domain in terms of the *y* variable, and is called the cell problem. Only in one dimension can one solve it in closed form; as a result, we know the homogenized coefficient explicitly. In several dimensions, the corresponding elliptic boundary value problem can be homogenized, but the homogenized coefficients are not known explicitly in general.

2.2 Reaction–Diffusion Traveling Fronts in Periodic Media

Now let us consider what happens if we let the reaction–diffusion (R-D) fronts discussed in Section 2.1 pass through a medium with periodic structure. If we model the medium with a periodic coefficient, then a model equation for R-D fronts is

$$u_t = (a(x)u_x)_x + f(u), \tag{2.10}$$

where $a(x)$ is a positive 1-periodic smooth function and $f(u)$ is a nonlinear function of one of the five types. Since we expect solutions to behave like fronts, we should see them in the large-space and large-time scaling limit. That is, let us consider (2.10) under the change of variables $x \to \varepsilon^{-1}x$, $t \to \varepsilon^{-1}t$, for ε small. The rescaled equation is

$$u_t^\varepsilon = \varepsilon(a(\varepsilon^{-1}x)u_x^\varepsilon)_x + \varepsilon^{-1}f(u^\varepsilon), \tag{2.11}$$

which resembles a homogenization problem except that there is also a singular prefactor ε^{-1} in front of the nonlinear term. We realize that there are two scales present in this problem. One is the width of the front, and the other is the wavelength of the periodic medium. The first one is easy to capture if we look at the rescaled form of a traveling front in a homogeneous medium, or $U(\varepsilon^{-1}(x - ct))$. The second one can be built in as in the homogenization ansatz (2.2). Combining the two ideas, we come up with the following two-scale ansatz for R-D fronts in periodic media:

$$u^\varepsilon \sim U(\varepsilon^{-1}(x - c^*t), \varepsilon^{-1}x) + \cdots, \tag{2.12}$$

where c^*, the average wave speed, plays the role of a^* in the homogenization example shown before. Certainly, we impose periodicity in $y = \varepsilon^{-1}x$, and a 0 or 1 far-field boundary condition in $s = (x - c^*t)/\varepsilon$.

Substituting (2.12) into (2.11), we find that U as a function of (s, y) satisfies the PDE

$$(\partial_s + \partial_y)(a(y)(\partial_s + \partial_y)U) + c^*U_s + f(U) = 0. \tag{2.13}$$

If (2.13) has a solution under the boundary conditions

$$U(s, \cdot) \text{ has period } 1, \quad U(+\infty, y) = 1, \quad U(-\infty, y) = 0, \tag{2.14}$$

the leading term of (2.12) is actually an exact solution! Recalling that the scaling was just to motivate ourselves, we see that we could have worked with the original equation (2.10) to begin with. The exact traveling front then has the functional form $U(x-ct,x)$, and it was first found and constructed in [240].

Comparing (2.2) and (2.12), we see that the two scales of (2.12) are not necessarily separate. In fact, they can be arbitrary, while in (2.2), the two scales are vastly separate. In this sense, (2.12) is a general two-scale representation. Also for this reason, we end up with a PDE cell problem to solve instead of an ODE cell problem. We will see that what makes (2.12) possible is the nonlinearity $f(U)$, and that the extreme cases when the front width is either much larger or much smaller than the wavelength of the medium are simpler.

It is easy to generalize the above form of traveling front to several spatial dimensions. Let us consider an R-D equation of the form

$$u_t = \nabla_x \cdot (a(x)\nabla_x u) + b(x) \cdot \nabla_x u + f(u), \quad u|_{t=0} = u_0(x), \qquad (2.15)$$

where

(A1): $a(x) = (a_{ij}(x))$, $x = (x_1, x_2, \ldots, x_n) \in \mathbb{R}^n$ is a smooth positive definite matrix on $\mathbb{R}^n$, 1-periodic in each coordinate x_i;

(A2): $b(x) = (b_j(x))$ is a smooth divergence-free vector field, 1-periodic in each coordinate x_i, with mean zero.

Equations of the form (2.15) appear in the study of premixed flame propagation through turbulent (random) media [56], where u is the temperature of the combustible fluid, $b(x)$ is the prescribed turbulent incompressible (divergence-free) fluid velocity field with zero ensemble mean, $f(u)$ is the Arrhenius reaction term, and $a(x)$ is taken as a constant matrix. Since the fluid velocity b is given as we solve for the temperature u, the above problem is called passive, and the traveling fronts are called passive fronts. In [56], formal asymptotic analysis suggests that u propagates with an averaged (effective) speed, also called the turbulent flame speed [56, 203]. Turbulence refers to complex random flows involving a wide range of spatial and temporal scales. Let us first consider periodic media to achieve a good preliminary understanding of effective flame speed. In Chapter 5, we shall give a definitive answer to front speeds of (2.15) in random (turbulent) media.

Let us fix a unit vector $k \in R^N$ and look for a traveling wave (front) moving in this direction with speed $c = c(k)$. The traveling front is of the form

$$u(x,t) = U(k \cdot x - ct, x), \qquad (2.16)$$

where the front speed c is an unknown constant depending on k, while U, the front profile, satisfies as a function of $s = k \cdot x - ct$ and $y = x$ the boundary conditions

$$U(-\infty, y) = 1, \quad U(+\infty, y) = 0, \quad U(s, \cdot) \text{ has period } 1. \qquad (2.17)$$

Upon substitution into equation (2.15), we obtain the following traveling-front equation for $U = U(s,y)$ and c:

$$(k\partial_s + \nabla_y)(a(y)(k\partial_s + \nabla_y)U) + b(y) \cdot (k\partial_s + \nabla_y)U + cU_s + f(U) = 0. \qquad (2.18)$$

The above form of traveling fronts (2.16) in periodic media and the mathematical study of (2.18) were initiated in the author's work in the early 1990s on bistable and ignition nonlinearities [240, 241, 242, 243], where existence and uniqueness are proved under suitable conditions.

A special case of (2.18) is when a is the identity, $b(y) = (b_1(y'),0)$, $y' = (y_2,\ldots,y_N)$, and $k = (1,0,\ldots,0)$. Such a vector field b is called shear flow. Then $u = U(x_1 - ct,x')$, $x' = (x_2,\ldots,x_N)$, and (2.18) reduces to

$$\Delta_{s,y'}U + (c + b_1(y'))U_s + f(U) = 0, \tag{2.19}$$

a semilinear elliptic PDE.

Equation (2.19) appeared earlier ([27] and references therein) as a model of flame propagation inside an infinite cylinder $(s,y') \in \mathbb{R} \times D$ for type-5 nonlinearity. The cylinder has a bounded cross section D, and the boundary condition on y' is zero Neumann, so the cylinder boundary is insulated for heat transfer. Existence and uniqueness of solutions to (2.19) is thoroughly studied for nonlinearities of types 1 through 5 in [29, 30, 31].

Interestingly, mathematicians were not alone in thinking about traveling fronts in periodic media. Theoretical biologists have long been interested in R-D fronts since the days of Fisher [91] and Hodgkin and Huxley [167]. An interdisciplinary problem of fundamental importance often draws attention and ideas from different scientific communities. Indeed, a different notion of traveling front in periodic media was proposed by biologists [221] in the mid 1980s. A traveling (pulsating) front is a solution $u(x,t)$ satisfying

$$
\begin{aligned}
u(x,t - L\cdot k/c) &= u(x+L,t), \quad \forall(x,t), \\
u(x,t) &\to 1 \text{ as } x\cdot k \to -\infty, \\
u(x,t) &\to 0 \text{ as } x\cdot k \to +\infty,
\end{aligned}
\tag{2.20}
$$

where L is the (vector) period of the media, c the front speed. The solution repeats itself in time $L\cdot k/c$ if it is observed at two points a distance L apart in space. Clearly, $u(x,t) = U(k\cdot x - ct,x)$ is such a front. In [221], formal arguments and linearizations at the unstable state $u = 0$ are made to find approximate solutions in one spatial dimension in the case of a (type-1) KPP reaction. However, error estimates of approximations are not demonstrated.

Interestingly in the late 1970s, about six years earlier than [221], mathematicians then working in the former Soviet Union had already developed a probabilistic functional integration method [94, 100] to find the KPP minimal speeds in periodic media of any dimensions. In the mid 1980s, this line of work was published in detail in the West [95, 96]. Though the work was quickly known in the mathematics community, apparently the authors of [221] were unaware of it, partly because of the lack of communication across scientific and geographical boundaries at the time.

Likewise, [240, 241, 242, 243] were done without knowledge of [221]. The analytical form (2.16)–(2.18) turns out to be more friendly to work with than a property of the time-dependent solution (2.20).

The probabilistic method [94, 100, 96] relies on the large-deviation technique to analyze the Feynman–Kac representation of KPP solutions. It leads to a variational formula for KPP minimal front speeds, and also serves as a rigorous justification of the formal linearization analysis [221]. In the physics literature, the method of linearization at an unstable state to determine front speeds is known as the marginal stability criterion (MSC) [210]. It originated in the 1950s from the plasma community [43] and was used by physicists in studying pattern selection in the early 1980s [63, 140]. Pattern selection refers to the dynamic selection of a front among a continuum of front solutions from a class of initial data. KPP is one example of a pattern-forming system in which dynamic selection is called for. In the case of homogeneous media, the works [8, 9] established the MSC of the KPP front speed $2\sqrt{f'(0)}$ by the PDE method.

The probabilistic method [100, 96] proved that MSC also holds for KPP in inhomogeneous media. We shall discuss this method in conjunction with periodic homogenization of HJ equations in the next section. Its advantage is that it bypasses the front profile and goes straight to the front speed. Impressively, it was worked out also for random media in one spatial dimension [100, 96]. PDE methods are more robust, and can handle more general forms of equations and nonlinearities, though they are traditionally restricted to deterministic media. We shall see in Chapter 5 that combining ideas of the large-deviation and PDE methods is a way to handle equation (2.15) in the random setting in arbitrary dimensions and to solve the turbulent front speed problem [203, 194] for KPP.

2.3 Existence of Traveling Waves and Front Propagation

Let us state the existence results for bistable and ignition fronts [240, 242].

Theorem 2.1. *Let T^n be the n-dimensional unit torus and $\|\cdot\|_{H^m(T^n)}$ the Sobolev norm of functions on T^n with up to m integrable derivatives. Define $\bar{a} = \int_{T^n} a(x)\,dx$, and assume that conditions (A1) and (A2) hold.*

1. *If the nonlinearity $f(U)$ is of type 3 (bistable nonlinearity) with $\mu \in (0, \frac{1}{2})$, there is a positive number δ_{cr} such that if $\|a(x) - \bar{a}\|_{H^m(T^n)} < \delta_{cr}$, $\|b(x)\|_{H^m(T^n)} < \delta_{cr}$, $m > n + 1$, then equation (2.18) has a unique classical solution (U, c) such that $0 < U < 1$, $U_s < 0$ for all $(s, y) \in \mathbb{R} \times T^n$, and $c > 0$.*
2. *If the nonlinearity $f(U)$ is of type 5 (combustion nonlinearity with ignition temperature), then for all a and b, equation (2.18) has a unique classical front (U, c) satisfying the same properties.*

Here uniqueness means that c is uniquely determined by the coefficients (a, b) and the nonlinearity $f(U)$, and U is unique up to a constant translation in s due to the translation-invariance of equation (2.18). The threshold phenomenon in the bistable case is because the unequal potential wells of the antiderivative of $f(u)$ (which are essentially the driving force behind front motion) can have effectively the same depth due to the influence of periodic media. Front speed is zero, and equation (2.15) has a stationary front solution $u = u(x)$. A similar situation occurs in the homogeneous case in which the intermediate zero of $f(u)$ is equal to $\frac{1}{2}$.

As in homogeneous media, type $(1,2,4)$ front speeds occupy an interval $[c_*, \infty)$, or the speed spectrum is a continuum. More precisely, we have [20] the following theorem.

Theorem 2.2. *If reaction nonlinearity f is of type $(1,2,4)$, there exists $c_* > 0$ such that no solution exists to (2.17)–(2.18) if $c < c_*$, and a monotone decreasing (in s) solution exists to (2.17) if $c \geq c_*$.*

Variational formulas of front speeds of type $(1,3,5)$ will be discussed later.

The next problem is to show that under certain conditions on the initial data, the time-dependent solutions behave like these special traveling-front solutions. Let us first state front propagation results for the bistable and ignition reaction [243].

Theorem 2.3 (Front Propagation). *Consider the initial value problem for equation (2.15) with initial data $0 \leq u_0(x) \leq 1$. Let f be of type 3 with $\mu \in \left(0, \frac{1}{2}\right)$ or of type 5 with $f'(1) < 0$. Assume in the context of type-3 nonlinearity that a traveling wave solution $U(k \cdot x - c(k)t, x)$ exists for every unit vector $k \in \mathbb{R}^n$. Let $s \in \mathbb{R}$ and let the plane orthogonal to k be $S = \{y \in \mathbb{R}^n | y = x - (k \cdot x)k, \quad \forall x \in \mathbb{R}^n\}$.*

I. *Suppose the initial date are frontlike: $u_0(x) \to 0$ sufficiently fast as $k_0 \cdot x \to -\infty$, and $u_0(x) \to 1$ sufficiently fast as $k_0 \cdot x \to -\infty$, uniformly in $S(k_0)$, for some $k_0 \in \mathbb{R}^n$. Then*

$$\lim_{t \to \infty} u(t, sk_0 t) = \begin{cases} 1, & s > c(k_0), \\ 0, & s < c(k_0). \end{cases}$$

II. *Suppose the initial data are pulselike: for some unit vector k, $u_0(x) \to 0$ sufficiently fast as $k_0 \cdot x \to -\infty$; $u_0(x) > \mu + \eta$, $|k \cdot x| < L$, for some positive constants η and L (θ replacing μ for f of type 5). Then there is a positive number $L_0(\eta) > 0$ such that if $L \geq L_0$, then*

$$\lim_{t \to \infty} u(t, skt) = \begin{cases} 1, & c(k) < s < -c(-k), \\ 0, & s < c(k) \text{ or } s > -c(-k). \end{cases}$$

The existence, uniqueness, and propagation results above are all based on maximum principles. The idea is to bound from above and below the exact solutions by simplified comparison functions, then extract asymptotic information. Let us explain the main ingredients below.

Consider equation (2.18) with a type-5 nonlinearity. Our first observation is that the three linear terms there do not form a strongly elliptic operator (such as the Laplacian $\Delta_{s,y}$), since the second derivatives are along directions

$$(k_i, 0, \ldots, 0, y_i, 0, \ldots, 0) \in \mathbb{R}^{n+1}, \quad i = 1, \ldots, n,$$

which do not cover all $n + 1$ directions. The other derivative along direction

$$(1, 0, \ldots, 0) \in \mathbb{R}^{n+1}$$

is the s derivative of U. Hence if c is not equal to zero, we have a parabolic operator (similar to the heat operator $\partial_t - \Delta_x$). This may sound like trouble, since for the standard heat equation, we cannot pose a boundary value problem in t.

However, what saves us is that the s direction of the infinite cylinder is not characteristic, since it is not orthogonal to all the directions $(k_i, 0, \ldots, 0, y_i, 0, \ldots, 0)$. The other observation is that (2.18) is translation-invariant in s. The loss of ellipticity is absent in the shear flow case, or equation (2.19).

Now, do we still have a strong maximum principle for the linear operator in (2.18),

$$Lu = (\nabla_y + k\partial_s)(a(y)(\nabla_y + k\partial_s)u) + b(y)^T \cdot (\nabla_y + k\partial_s)u + cu_s, \qquad (2.21)$$

even though it is not strongly elliptic?

As long as $c \neq 0$, the answer is yes, thanks to the parabolic maximum principle and the periodicity in y. Periodicity helps us to overcome the degeneracy! For classical maximum principles, we refer to [198, 226].

Now let us take $c = -1$ for convenience and prove the following result.

Proposition 2.4. *Let u be a classical solution of the differential inequality $Lu \leq 0$ ($Lu \geq 0$) on $\mathbb{R} \times T^n$. If u achieves its minimum (maximum) at (s_0, y_0) with s_0 finite, then $u \equiv constant$.*

Proof. We first treat the special case $n = 1, k = 1$, in which case we have

$$Lu = (\partial_s + \partial_y)(a(y)(\partial_s + \partial_y)u) + b(y)(\partial_s + \partial_y)u - u_s.$$

For the time being, unfold T into $\mathbb{R}$ and regard L as an operator on $\mathbb{R}^2$. If we make the change of variables

$$s' = \frac{1}{\sqrt{2}}(s - y), \quad y' = \frac{1}{\sqrt{2}}(s + y),$$

then

$$\partial_s = \frac{1}{\sqrt{2}}(\partial_{s'} + \partial_{y'}), \quad \partial_y = \frac{1}{\sqrt{2}}(-\partial_{s'} + \partial_{y'}), \quad \partial_s + \partial_y = \sqrt{2}\partial_{y'}.$$

In terms of (s', y'), Lu becomes

$$Lu = 2(au_{y'})_{y'} - \frac{1}{\sqrt{2}}u_{s'} + \left(\sqrt{2}b - \frac{1}{\sqrt{2}}\right)u_{y'}.$$

Here L is a standard parabolic operator in (s', y'), elliptic in y', and parabolic in s'. By the strong maximum principle for parabolic operators, we see that if u attains its minimum at some finite point (s'_0, y'_0), then

$$u \equiv constant \text{ if } s' \leq s'_0,$$

or
$$u \equiv \text{constant if } s - y \le s_0 - y_0.$$

By the periodicity of u in y, we see that $u \equiv$ constant for all s and y. If $n \ge 2$, we can always subject y to an orthogonal transform, i.e., $y = Qy'$, and then Lu becomes

$$Lu = (k\partial_s + Q^T\nabla_{y'})^T a(k\partial_s + Q^T\nabla_{y'})u + b^T \cdot (k\partial_s + Q^T\nabla_{y'})u - u_s$$
$$= (Qk\partial_s + \nabla_{y'})^T QaQ^T(Qk\partial_s + \nabla_{y'})u + b^T \cdot Q^T(Qk\partial_s + \nabla_{y'})u - u_s.$$

Choosing Q such that $Qk = e_1 = (1,0,\ldots,0)$ and setting $a_1 = QaQ^T$ and $b_1 = Qb$, we have

$$Lu = (e_1\partial_s + \nabla_{y'})^T a_1(e_1\partial_s + \nabla_{y'})u + b_1^T \cdot (e_1\partial_s + \nabla_{y'})u - u_s.$$

If we make the change of variables

$$s' = \frac{1}{\sqrt{2}}(s - y_1'), \quad z_1 = \frac{1}{\sqrt{2}}(s + y_1'), \quad z_i = \frac{1}{\sqrt{2}}y_i', \ i \ge 2,$$

then just as in the case $n = 1$, we have

$$Lu = 2\nabla_z^T(a_1\nabla_z u) - \frac{1}{\sqrt{2}}u_{s'} + \sqrt{2}b_1^T \cdot \nabla_z u - \frac{1}{\sqrt{2}}u_{z_1}.$$

By the strong maximum principle for parabolic operators, if u attains its minimum at some finite point $P_0 = (s_0', z_0)$, then

$$u = \text{constant if } s' \le s_0',$$

or
$$u = \text{constant if } s - y_1' \le s_0 - y_{1,0}'.$$

In terms of (s,y), this asserts that u is a constant under some hyperplane that is not orthogonal to the s-axis. The periodicity of u in y implies that $u \equiv$ constant for all s and y. The proof is complete. $\square$

Let us outline the two steps of the construction for existence of type-5 solutions based on a degree-theoretic approach. In step one, we consider a family of elliptically regularized problems ($\varepsilon > 0$, $\tau \in [0,1]$),

$$\varepsilon U_{ss} + L_\tau U + \tau f(U) = 0, \quad (s,y) \in \Omega_a = [-a,a] \times T^n, \qquad (2.22)$$

subject to the boundary conditions $U(-a,y) = 1$, $U(+a,y) = 0$. The operator L_τ is L with a replaced by $\langle a \rangle(1 - \tau) + \tau a$ and b replaced by τb, with $\langle \cdot \rangle$ being the period average.

To remove the translation-invariance of solutions, we must also impose a normalization condition: $\max_{y \in T^n} U(0,y) = \theta$. By the elliptic maximum principle, we know that U is bounded between 0 and 1 and that $U_s < 0$. Elliptic regularity also tells us that the maximum of ∇U is bounded independently of a and τ. The parameter τ links the linear problem ($\tau = 0$) with the problem of interest $\tau = 1$.

Consider the space $E = C^1(\Omega_a) \times R$. For $(v,c) \in E$, $\tau \in [0,1]$, let $u = \varphi_\tau(v,c)$ be the unique solution of the elliptic boundary value problem

$$\varepsilon u_{ss} + L_\tau u + \tau f(v) = 0$$

under the same 0 and 1 boundary conditions. Define

$$h_\tau(v,c) = \max_{\substack{y \in T^n \\ s=0}} \varphi_\tau(v,c).$$

Then the solution of (2.22) satisfies

$$u = \varphi_1(u,c), \quad h_1(u,c) = \theta. \tag{2.23}$$

Define $F_\tau(u,c) = (\varphi_\tau(u,c), c - h_\tau(u,c) + \theta)$, $\tau \in [0,1]$. Now the existence of the solution is the same as the fixed-point problem

$$F_1(u,c) = (u,c).$$

Notice that the mapping $(\tau,(u,c)) \to F_\tau(u,c)$ from $[0,1] \times E$ to E is continuous and compact. Due to the a priori bounds on the solutions and their derivatives, the Leray–Schauder degree of the mapping $\mathrm{Id} - F$ is well defined on a bounded closed set of the form

$$D \equiv \{(u,c) \in E, \|u\|_{C^1(R_a)} \leq K, \ |c| \leq K\},$$

where K some constant larger than the bounds of the solutions. This is because the zeros of $\mathrm{Id} - F$ cannot occur on the boundary of the set D. The degree is a measure of the number of zeros counting multiplicity, and is invariant under a change of $\tau \in [0,1]$; see [254] for details. If the degree is nonzero, then we have a fixed point. This is easily checked when $\tau = 0$, since (2.23) is explicitly solvable, and we find that the degree is equal to one.

In step two, we pass to the limit $a \to \infty$ first and then to the limit $\varepsilon \to 0$. To this end, the main technical work is to bound the wave speed c away from 0 and ∞ independently of both parameters. This can be achieved with the help of comparison principles of wave speeds for the $a \to \infty$ limit, see [241], and the identity $c = -\int_{R \times T^n} f(U)$ for the $\varepsilon \to 0$ limit; see [242].

Thanks to the normalization condition and $U_s \leq 0$, we have $U \leq \theta$ if $s \geq 0$. Hence we have a linear equation for U on $s \geq 0$. We can now look for a special decay solution of the form $\bar{U} = e^{\mu s} \psi(y)$ with $\psi(y) > 0$ and $\mu < 0$. This decay solution has a continuous limit as $\varepsilon \to 0$ along a subsequence, and $\limsup_{\varepsilon \to 0} \mu < 0$. It follows that the limiting solution must decay to zero as $s \to +\infty$. As $s \to -\infty$, monotonicity implies $U(s,y) \to U_-$. It is not hard to show that U_- satisfies the elliptic equation (dropping s derivatives from (2.18)

$$\nabla_y \cdot (a(y)\nabla_y U) + b(y) \cdot \nabla_y U + f(U) = 0 \tag{2.24}$$

under periodic boundary conditions. Since $f(U) \geq 0$, the maximum principle implies that (2.24) has only constant nonnegative solutions. Thus U_- equals either θ or 1. In the former case, $U \leq \theta$, and hence $f(U) \equiv 0$, for any (s,y). So $LU = 0$, for all (s,y), and thus U attains its maximum θ at a finite point $(0, y^*)$ as a result of imposing a normalization condition at $s = 0$.

By the strong maximum principle property of the operator L in Proposition 2.4, U must be identically equal to a constant, which is impossible since it has a zero limit at $s = +\infty$. We have constructed a desired traveling-front solution with the property $U_s < 0$ (strict inequality again follows from the strong maximum principle of L).

The other bonus of the strong maximum principle of L is that the sliding domain method [30, 145] applies to show that traveling-front solutions to (2.18) must be unique. The uniqueness means that there is only one value of the wave speed c for any given coefficients (a,b) and nonlinearity f of type 5. Moreover, the profile U is unique up to a constant translation in s, and is strictly monotone in s.

The basic argument to show monotonicity is as follows. First, we compare $U(s,y)$ and its translate $U_\lambda = U(s - \lambda, y)$. For large λ, U_λ is larger than U for those points (s,y) in a bounded cylinder. The bounded cylinder is large enough that $U(s,y)$ is close to either 0 or 1 outside of it. Then $w_\lambda \equiv U(s - \lambda, y) - U(s,y)$ satisfies the differential inequality $Lw_\lambda \leq 0$ outside of the finite cylinder. The strong maximum principle for L implies that $w_\lambda > 0$ holds at any point. Then we decrease λ to the infimum value λ_0 at which U_λ is no less than U. Now $w_{\lambda_0} \geq 0$.

Again, the strong maximum principle implies that at λ_0, U and U_λ must be identical, which is possible only if $\lambda_0 = 0$, due to the front boundary conditions. We conclude that U is strictly monotone and actually has positive derivative anywhere (by invoking the minimum principle on the derivative). A similar argument can be carried out for any two profiles to show that they agree up to a constant translation, and for the uniqueness of c, see [241].

The other nonnegative f of type $(1,2,4,5)$ can be approximated by a sequence of ignition nonlinear functions f_{θ_n}, where $\theta_n \downarrow 0$. Let χ_θ $(\theta < \frac{1}{2})$ be a smooth compactly supported function such that $\chi_\theta(u) = 0$ if $u \leq \theta$, and $\chi_\theta(u) = 1$ if $u \geq 2\theta$. Defining $f_\theta = \chi_\theta f$ will do. The speed c_θ is monotone in θ. One may then pass the approximate type-5 solutions to the limit and verify that the limit remains a front solution. The minimal speed c_* is equal to $\lim_{n \to \infty} c_{\theta_n}$. One then uses the corresponding solution U_* to prove the existence of other solutions for $c > c^*$, and the converging sequence $(U_{\theta_n}, c_{\theta_n})$ to exclude solutions for $c < c_*$; see [20] for complete proofs. The KPP minimal speed is related to the existence of a positive solution of the form $e^{\lambda s} \psi(y)$ to the linearized equation at $U = 0$, which will be presented later in a more general context.

For type-3 nonlinearity, new difficulties come up in step two. Because f changes sign, there can be many nontrivial periodic solutions of (2.24). As we know, traveling fronts may not exist for all $a(y)$ due to the existence of steady states. The convenient method for establishing traveling waves in type 3 (the bistable case) is to use the method of continuation [242, 244] and treat a family of problems in which $a(y)$ is replaced by $(1 - \delta)\langle a \rangle + \delta a(y)$.

We start with δ small and obtain solutions by perturbing the known one-dimensional front. The linearized operator has a simple eigenvalue at zero, and the rest of the spectrum is isolated away from zero. The monotonicity of the perturbed solutions guarantees that the same spectral property of the linearized operator remains, and so the perturbation continues on δ. Since each perturbative step relies on the contraction mapping principle, there is no difficulty as $|s| \to \infty$. Of course, the same difficulty arises if we want to show that the continuation goes to any value of $\delta \in [0, 1]$, which we know is false in general.

The continuation method is convenient in that it deals with the problem on the infinite domain, where estimates of solutions are usually simpler. However, it relies on good spectral properties of the linearized operators. It works for nonlinearity of type 3, also type 5 if $f'(1) < 0$, as well as spatially periodic conservation laws [244]. For type 5, the assumption $f'(1) < 0$ can be removed by further approximation of nonlinearity. To summarize, the degree-theoretic method and the continuation method with the help of maximum principles guarantee the existence of traveling-front solutions as stated.

Let us sketch the proof of statement (I) of Theorem 2.3 in the case of f of type 5, and refer to [243] for the complete proof. The proof is similar for type 3. The idea is to construct subsolutions (supersolutions) using the parabolic maximum principle [198, 226] and the traveling wave solutions. The long-time asymptotics of the subsolutions (supersolutions) rely on the decay property of solutions of the variable-coefficient linear parabolic equations of the form

$$u_t = \nabla \cdot (a(x)\nabla u) + b(x) \cdot \nabla u, \quad \nabla \cdot b(x) = 0. \tag{2.25}$$

The fundamental solution (Green's function) of (2.25), in turn, has pointwise lower and upper bounds in terms of heat kernels [169, 84, 186].

First we note that due to the fast convergence of u_0 to 1 as $k \cdot x \to \infty$, there are a number $\xi_0 > 0$ large enough and a positive spatially decaying function $q_0 = q_0(k \cdot x) < (1 - \theta)/2$ such that

$$U(k \cdot x - \xi_0, x) - q_0(k \cdot x) \leq u_0(x)$$

on $\mathbb{R}^n$. Now consider the function

$$u_l \equiv U(k \cdot x - c(k)t - \xi_1(t), x) - q_1(t, x),$$

where ξ_1 and q_1 will be chosen to satisfy

$$\xi_1'(t) > 0, \quad \xi_1(t) > 0, \quad \xi_1(t) = o(t), \quad t \to \infty.$$

We calculate

$$\begin{aligned} N[u_l] &= u_{l,t} - \nabla_x \cdot (a(x)\nabla_x u_l) - b(x) \cdot \nabla_x u_l - f(u_l) \tag{2.26} \\ &= -\xi_1'(t)U_s - q_{1,t} + \nabla_x \cdot (a(x)\nabla_x q_1) + b(x) \cdot \nabla_x q_1 + f(U) - f(U - q_1). \end{aligned}$$

There exists $\delta \in (0, \theta)$ sufficiently small that if $q \in \left[0, \frac{1-\theta}{2}\right]$ and $U \in [1 - \delta, 1]$, then

$$f(U) \leq f(U - q).$$

Since $0 \leq q \leq q_0 < \frac{1-\theta}{2}$, we have for $U \in [1 - \delta, 1]$,

$$N[u_l] \leq -\xi_1'(t)U_s - q_{1,t} + \nabla_x \cdot (a(x)\nabla_x q_1) + b(x) \cdot \nabla_x q_1. \tag{2.27}$$

If $U \in [0, \delta]$, then $f(U) = f(U - q_1) = 0$, so (2.27) holds with an equality sign. If $U \in (\delta, 1 - \delta)$, then there exists $\beta > 0$ such that $U_s \geq \beta$ and $|f(U) - f(U - q_1)| \leq Kq_1$ for some $K > 0$. It follows that

$$N[u_l] \leq -\xi_1'\beta - q_{1,t} + \nabla_x \cdot (a(x)\nabla_x q_1) + b(x) \cdot \nabla_x q_1 + Kq_1. \tag{2.28}$$

Let us choose q_1 to satisfy the equation

$$q_{1,t} = \nabla_x \cdot (a(x)\nabla_x q_1) + b(x) \cdot \nabla_x q_1, \quad q_1|_{t=0} = q_0(k \cdot x). \tag{2.29}$$

To make u_l a subsolution or $N[u_l] \leq 0$, we just need to impose the condition

$$-\xi_1'\beta + Kq_1 \leq 0, \quad \text{or} \quad -\xi_1'\beta + K\|q_1\|_{L^\infty(\mathbb{R}^n)} = 0,$$

or

$$\xi_1' = \frac{K\|q_1\|_{L^\infty(\mathbb{R}^n)}}{\beta} > 0, \tag{2.30}$$

with $\xi_1(0) = \xi_0 > 0$. By our early comments on the fundamental solution of (2.29), $\|q_1\|_{L^\infty} = o(1)$ as $t \to \infty$. Therefore $\xi_1(t) = o(t)$. We have shown that u_l is a sub-solution, $u_l \leq u$. A supersolution can be constructed in a similar way. We conclude that statement (I) holds.

2.4 KPP Fronts and Periodic Homogenization of HJ Equations

Consider the KPP front $u(x,t) = U(k \cdot x - c_*(k)t, x)$. Then under the hyperbolic scaling $t \to \varepsilon^{-1}t$, $x \to \varepsilon^{-1}x$, the scaled solution $u^\varepsilon(x,t) = U((k \cdot x - c_*(k)t)/\varepsilon, x/\varepsilon)$ converges to a step function traveling at speed $c^*(k)$ in the direction k. This way of finding the speed by scaling limit can be done directly on the PDE. To fix ideas, let us consider the homogeneous medium in one spatial dimension. The scaled equation is

$$u_t^\varepsilon = \frac{\varepsilon}{2}u_{xx}^\varepsilon + \varepsilon^{-1}f(u^\varepsilon), \tag{2.31}$$

where we have modified the diffusion constant to $\frac{1}{2}$ for convenience of stochastic representation. The initial condition is the indicator function $1_{\overline{G_0}}(x)$, where G_0 is an open interval. Our goal is to recover $c_* = \sqrt{2f'(0)}$ from (2.31).

Let $c(u) = u^{-1} f(u)$. Then equation (2.31) can be regarded as a heat equation with a time-dependent potential c. The solution has a well-known stochastic representation formula and the Feynman–Kac formula [96, Chapter 2], [55, Chapter 3]

$$u^{\varepsilon}(x,t) = E_x g(X_t^{\varepsilon}) \exp\left\{\varepsilon^{-1} \int_0^t c(u(t-s, X_s^{\varepsilon})) \, ds\right\}, \qquad (2.32)$$

where $X_t^{\varepsilon} = x + \sqrt{\varepsilon} W_t$, where W_t is the standard Wiener process. Since $0 < u^{\varepsilon} \leq 1$, it follows from the KPP assumption of f that

$$u^{\varepsilon}(x,t) \leq E_x g(X_t^{\varepsilon}) \exp\left\{\varepsilon^{-1} \int_0^t c(0) \, ds\right\} = e^{f'(0)t/\varepsilon} P(X_t^{\varepsilon} \in \overline{G_0}). \qquad (2.33)$$

If we denote the distribution of $\sqrt{\varepsilon} W_t$ by P_{ε}, then it follows from the properties of the standard Wiener process that $P_{\varepsilon} \overset{\text{law}}{\to} \delta_x$, where δ_x is the measure with unit mass concentrated at the function identically equal to x. The covariance of $X^{\varepsilon} - x$ equals $\varepsilon \min(s,t)$. There is, however, an exponentially small probability that X^{ε} may escape from being close to x. These are called rare events. The large-deviation theory [229] studies the asymptotics of such small probabilities. A family of probability measures P_{ε} on a complete separable metric space X is said to obey the large-deviation principle (LDP) with a rate function $I(\cdot)$ if there exists a function $I(\cdot) : X \to [0, \infty]$ satisfying:

1. $0 \leq I(x) \leq \infty, \, \forall x \in X$.
2. $I(\cdot)$ is lower semicontinuous.
3. For each $l < \infty$, the set $\{x : I(x) \leq l\}$ is compact in X.
4. For each closed set $C \subset X$,

$$\limsup_{\varepsilon \to 0} \varepsilon \log P_{\varepsilon}(C) \leq - \inf_{x \in C} I(x). \qquad (2.34)$$

5. For each open set $G \subset X$,

$$\liminf_{\varepsilon \to 0} \varepsilon \log P_{\varepsilon}(G) \geq - \inf_{x \in G} I(x). \qquad (2.35)$$

If $x = 0$, the scaled Wiener process $X^{\varepsilon}(s)$, $s \in [0,1]$, satisfies LDP with the rate function $I = I(g)$ defined for any continuous function g on $[0,1]$ with $g(0) = 0$ as

$$I(g) = \frac{1}{2} \int_0^1 (g'(s))^2 \, ds \qquad (2.36)$$

if $g(s)$ is absolutely continuous with L^2 derivative, otherwise $I(g) = \infty$. See [229, Section 5] for a proof. In view of (2.36), we have

$$\limsup_{\varepsilon \to 0} \varepsilon \log P(X_t^{\varepsilon} \in \overline{G_0}) = \limsup_{\varepsilon \to 0} \varepsilon \log P_{\varepsilon}(\overline{G_0}) = - \inf_{\substack{\varphi_0 = x \\ \varphi_t \in \overline{G_0}}} \int_0^t |\dot{\varphi}(s)|^2 \, ds, \qquad (2.37)$$

which is a minimal action (cost) equal to $-d^2(x, G_0)/2t$, where d is the distance function. It follows from (2.33) that

$$\lim_{\varepsilon \to 0} \varepsilon \log u^\varepsilon(x,t) \le f'(0)t - \frac{d^2(x, G_0)}{2t} \equiv V. \tag{2.38}$$

Clearly,

$$\lim_{\varepsilon \to 0} u^\varepsilon(x,t) = 0 \quad \forall (x,t) \in N \equiv \{(x,t) : V(x,t) < 0\}. \tag{2.39}$$

The function $V(x,t)$ is continuous, and the convergence is uniform on compact subsets. Setting $V(x,t) = 0$ gives the front equation $d(x, G_0) = \sqrt{2f'(0)}t$ and the desired front speed $c_* = \sqrt{2f'(0)}$. One verifies by direct calculation that the function $V(x,t)$ satisfies the HJ equation

$$\psi_t - \psi_x^2/2 - f'(0) = 0, \tag{2.40}$$

and initial data $\psi(x,0) = 0$ if $x \in \overline{G_0}$, $\psi(x,0) = -\infty$ otherwise.

It remains to show that $u^\varepsilon \to 1$ if $V(x,t) > 0$, or that $u^\varepsilon(x,t) \ge 1 - \lambda$, on any compact subset of $P = \{(x,t) : V(x,t) > 0\}$ for any small positive number λ. We need a more general Feynman–Kac formula with stopping times installed [96]:

$$u^\varepsilon(x,t) = E_{t,x} u^\varepsilon(t_\tau, X_\tau^\varepsilon) \exp\left\{ \varepsilon^{-1} \int_0^\tau c(u^\varepsilon(t-s, X_s^\varepsilon)) \, ds \right\}$$

$$= E_{t,x} 1_{\tau=\tau_1} u^\varepsilon(t_{\tau_1}, X_{\tau_1}^\varepsilon) \exp\left\{ \varepsilon^{-1} \int_0^{\tau_1} c(u^\varepsilon(t-s, X_s^\varepsilon)) \, ds \right\}$$

$$+ E_{t,x} 1_{\tau=\tau_2} u^\varepsilon(t_{\tau_2}, X_{\tau_2}^\varepsilon) \exp\left\{ \varepsilon^{-1} \int_0^{\tau_2} c(u^\varepsilon(t-s, X_s^\varepsilon)) \, ds \right\}, \tag{2.41}$$

where τ_1, τ_2, and τ are given by

$$\tau_1 = \inf\left\{ s : u^\varepsilon(t-s, X_s^\varepsilon) \ge 1 - \lambda \right\},$$
$$\tau_2 = \inf\left\{ s : V(t-s, X_s^\varepsilon) = 0 \right\},$$
$$\tau = \min(\tau_1, \tau_2).$$

A stopping time is a random variable whose value depends only on the information known up to this value of time, also called the hitting time (the first time that some event occurs). The minimum of two stopping times is also a stopping time. See [72, Section 3.1] for more examples. The expectation takes a double subscript to mean that it acts on the vector trajectory $(t-s, X_s^\varepsilon)$. Since $c \ge 0$, the first term on the right-hand side of (2.41) is bounded from below by

$$(1 - \lambda) E_{t,x}^\varepsilon 1_{\tau=\tau_1} = (1 - \lambda) P_{t,x}^\varepsilon(\tau = \tau_1).$$

For the second expectation term in (2.41), we need to control u^ε on $V = 0$ so that it is not too small and the exponential of the integral can balance it out. Then

$$u^\varepsilon \ge (1 - \lambda) P_{t,x}^\varepsilon(\tau = \tau_1) + P_{t,x}^\varepsilon(\tau = \tau_2) \ge 1 - \lambda,$$

and we would be done.

Note that over $s \in [0, \tau_2]$, we have $u^\varepsilon \in (0, 1 - \lambda]$, so $c(u^\varepsilon) \geq \min_{u \in [0, \lambda]} c(u) \equiv c_\lambda > 0$, and the exponential term indeed provides growth of order $O(\exp\{hc_\lambda/\varepsilon\})$, where $h \leq \tau_2$ is a positive number almost surely independent of ε. This is because it takes a positive amount of time for $((t - s), X_s^\varepsilon)$ to leave (t, x) where $V > 0$ to reach a point where $V = 0$. A lower bound of u^ε on the interface $V = 0$ of order $O(\exp\{-\delta/\varepsilon\})$ for small δ suffices. This lower bound requires $V(x, t)$ to satisfy a condition that for $(x, t) \in N$,

$$V(x,t) = \sup\left\{ f'(0)t - \int_0^t |\dot{\varphi}(s)|^2 \, ds : \varphi_0 = x, \varphi_t \in \overline{G_0}, (t - s, \varphi_s) \in N, s \in (0, t) \right\},$$
(2.42)

for any $t > 0$. Condition (2.42) says that $V(x, t)$ is the supremum of the action functional over the paths in the region of $V < 0$. The lower bound of u^ε on $V = 0$ is obtained by conditioning the stochastic path X^ε in formula (2.32) near the optimal path in region $V < 0$. The probability of the conditioning is controlled by the action function in (2.42), whose value over the optimal path is close to zero and so can be bounded from below by $-\delta$. The lower bound on u^ε on $V = 0$ of the form $O(\exp\{-\delta/\varepsilon\})$ holds.

Condition (2.42) is valid for the function V in (2.38). Provided that (2.42) continues to hold, the above argument extends to slowly varying media in higher dimensions, for example when $f = f(x, u) > 0$ for $u \in (0, 1)$, $f(u, x) < 0$ for $u < 0$ and $u > 1$, $f_u(x, 0) = \sup_{0 < u \leq 1} u^{-1} f(u, x)$. See [96] for complete results. The large-deviation method motivates the ansatz

$$u^\varepsilon \sim \exp\left\{ \frac{-I(x,t)}{\varepsilon} \right\},$$
(2.43)

and the PDE approach [81, 82, 83] based on the logarithmic change of variable $v^\varepsilon = -\varepsilon \ln u^\varepsilon$. Let $f(u) = u(1 - u)$. Then the function v^ε satisfies the equation

$$v_t^\varepsilon = \frac{\varepsilon}{2} v_{xx}^\varepsilon - \frac{1}{2} |v_x^\varepsilon|^2 + \exp\left\{ -\frac{v^\varepsilon}{\varepsilon} \right\} - f'(0),$$

where

$$v^\varepsilon(x, 0) = 0, \quad x \in G_0; \quad v^\varepsilon(x, t) \to +\infty, \quad \text{as } t \downarrow 0^+, \quad x \in G_0^c.$$
(2.44)

The next step is to pass to the limit $\varepsilon \to 0$ for v^ε. Comparison functions and maximum principles imply that the supremum norm and the Hölder norms (with exponent $\alpha \in (0, 1)$) of v^ε are bounded in any space–time compact set. Hence v^ε has a uniformly convergent subsequence with limiting function v.

The function v satisfies the variational inequality

$$\min\left[v_t + |v_x|^2/2 + f'(0), v \right] = 0, \quad x \in \mathbb{R}, \ t > 0.$$
(2.45)

This is understood as follows. Fix $T > 0$. If $v \geq 0$, then

$$v_t + |v_x|^2/2 + f'(0) \geq 0, \quad (x,t) \in \mathbb{R} \times (0,T], \qquad (2.46)$$

and on the set $\{v > 0\} \cap \mathbb{R} \times (0,T]$,

$$v_t + |v_x|^2/2 + f'(0) = 0, \qquad (2.47)$$

both in the viscosity sense. The viscosity sense in (2.46) means that for each smooth function φ, if $u - \varphi$ has a local minimum at $(x_0, t_0) \in \mathbb{R} \times (0,T]$, then

$$\varphi_t(x_0, t_0) + \frac{1}{2}|\varphi_x(x_0, t_0)|^2 + f'(0) \geq 0. \qquad (2.48)$$

In (2.47), we have in addition that if $v - \varphi$ has a local maximum at $(x_1, t_1) \in \mathbb{R}^n \times (0,T]$ and if $v(x_1, t_1) > 0$, then

$$\varphi_t(x_1, t_1) + \frac{1}{2}|\varphi_x(x_1, t_1)|^2 + f'(0) \leq 0. \qquad (2.49)$$

To see (2.45), we know that $v \geq 0$ by the maximum principle. Equation (2.44) implies the inequality

$$v_t^\varepsilon - \frac{\varepsilon}{2}v_{xx}^\varepsilon + \frac{1}{2}|v_x^\varepsilon|^2 + f'(0) \geq 0,$$

which yields (2.46) as $\varepsilon \to 0$ in the viscosity sense. Also on any compact subset of $\{v > 0\}$, $b\exp\{-\varepsilon^{-1}v^\varepsilon\} \to 0$; hence we have (2.47). The solution to equation (2.47) differs from that of (2.40) by a sign, and the solution to (2.45) can be written as $v = \max(-V, 0)$. In more general slowly varying media, the solution v in the variational inequality (2.45) with initial condition $v = 0$ on G_0, $v = +\infty$ on G_0^c admits a representation in terms of a two-player, zero-sum differential game with stopping times [83, 92], which resembles the action functional in (2.42).

We see from the above analysis that the HJ equation (2.47) or (2.40) carries the information on the KPP front speed. Let us exploit this KPP–HJ connection further and consider KPP fronts with minimal speed in periodic media by studying the equation

$$u_t = \frac{1}{2}\sum_{i,j=1}^n a_{ij}(x)u_{x_i x_j} + \sum_{i=1}^n b_i(x)u_{x_i} + f(x,u), \qquad (2.50)$$

with KPP nonlinearity $f(x,u)$ and initial data $g(x)$ of compact support G_0. The problem is solved in [100, 96] by the large-deviation method and a path integral representation of solutions as we illustrated above.

Again, the nonlinearity $f(x,u)$ can be approximated by $c(x) = f_u(x,0)$ times u, so that the implicit solution formula becomes explicit, and the large-deviation method yields the long-time front speed. Here we state the result [96].

Theorem 2.5. *Let $z \in \mathbb{R}^n$. Define the operator*

$$L_z = \frac{1}{2} \sum_{i,j=1}^{n} a_{ij}(y)(\partial_{y_i} - z_i)(\partial_{y_j} - z_j) + \sum_{i=1}^{n} b_i(\partial_{y_i} - z_i) + c(y) \tag{2.51}$$

on 1-*periodic functions in* $y \in T^n$, *the n-dimensional unit torus. Let* $\lambda = \lambda(z)$ *be the principal eigenvalue of* L_z, *which can be shown to be convex and differentiable in z. Let* $H(y)$ *be the Legendre transform of* λ,

$$H(y) = \sup_{z \in \mathbb{R}^n} [(y,z) - \lambda(z)],$$

$y \in \mathbb{R}^n$. *The function* $H(y)$ *is also convex and differentiable. Then for any closed* $F \subseteq \{y : H(y) > 0\}$, *we have* $\lim_{t \to \infty} u(t, ty) = 0$ *uniformly in* $y \in F$. *For any compact* $K \subseteq \{y : H(y) < 0\}$, *we have* $\lim_{t \to \infty} u(t, ty) = 1$ *uniformly in* $y \in K$.

It follows that the asymptotic front speed $v = v(e)$ along the unit direction e satisfies $H(ve) = 0$. If $\min_{\mathbb{R}^n} \lambda(z) > 0$, then the H equation can be solved to yield the *KPP speed variational formula*

$$v = v(e) = \inf_{(e,z)>0} \frac{\lambda(z)}{(e,z)}. \tag{2.52}$$

In fact, $\lambda(z)$ grows quadratically in z, and so the supremum in the definition of $H(y)$ is achieved. There exists z^* such that

$$0 = H(ve) = v(e, z^*) - \lambda(z^*),$$

and $(e, z^*) > 0$ due to $\lambda(z^*) > 0$. It follows that $v = \lambda(z^*)/(e, z^*) > 0$ and $\lambda(z)(v(e, z^*) - \lambda(z^*)) = 0 \geq \lambda(z^*)(v(e, z) - \lambda(z))$, implying

$$\frac{\lambda(z)}{(e,z)} \geq \frac{\lambda(z^*)}{(e,z^*)}.$$

This implies formula (2.52). The assumption $\min_{\mathbb{R}^n} \lambda(z) > 0$ holds if the operator L is self-adjoint or of the form $L = \nabla \cdot (a(x)\nabla \cdot) + b(x) \cdot \nabla \cdot$, where b is a mean-zero incompressible velocity.

Instead of going through the large-deviation method, let us follow the spirit of the logarithmic transform in the PDE approach and derive the same result. First consider equation (2.50) under the scaling $x \to \varepsilon^{-1} x$, $t \to \varepsilon^{-1} t$. The rescaled equation reads

$$u_t^\varepsilon = \frac{1}{2} \varepsilon \sum_{i,j=1}^{n} a_{ij}(\varepsilon^{-1} x) u_{x_i, x_j}^\varepsilon + \sum_{i=1}^{n} b_i(\varepsilon^{-1} x) u_{x_i}^\varepsilon + \varepsilon^{-1} f(\varepsilon^{-1} x, u^\varepsilon), \tag{2.53}$$

for which we make the change of variable

$$u^\varepsilon = \exp\{\varepsilon^{-1} v^\varepsilon\}. \tag{2.54}$$

Then v^ε satisfies the equation

$$v_t^\varepsilon = \frac{\varepsilon}{2}\sum_{i,j=1}^{n} a_{ij}(\varepsilon^{-1}x)v_{x_i,x_j}^\varepsilon + \frac{1}{2}\sum_{i,j=1}^{n} a_{ij}(\varepsilon^{-1}x)v_{x_i}^\varepsilon v_{x_j}^\varepsilon + \sum_{i=1}^{n} b_i(\varepsilon^{-1}x)v_{x_i}^\varepsilon + \frac{f(\varepsilon^{-1}x, u^\varepsilon)}{u^\varepsilon}.$$

$$(2.55)$$

The last term is bounded from above by $c(\varepsilon^{-1}x) = f_u(\varepsilon^{-1}x, 0)$, which also happens to be the right approximation of the nonlinearity for small values of u^ε. For locating the front or the region where u^ε is near zero, one can replace the nonlinear term by its linearization at u^ε equal to zero as we approach the front from the interior where $v^\varepsilon < 0$. Then equation (2.55) becomes the periodic homogenization problem of a viscous Hamilton–Jacobi equation.

The periodic homogenization of the inviscid Hamilton–Jacobi equation was first studied in [148]. Let v^ε be a solution of

$$v_t^\varepsilon + H(\nabla v^\varepsilon, \varepsilon^{-1}x) = 0, \quad x \in \mathbb{R}^n \times (0, +\infty), \qquad (2.56)$$

with initial data $v^\varepsilon(x, 0) = v_0$, where H is periodic in the second variable, say with period 1. Under the conditions that H is locally Lipschitz in all variables, $H(p, x) \to +\infty$ as $|p| \to +\infty$ uniformly in $x \in \mathbb{R}^n$, u_0 is bounded and uniformly continuous, and $\nabla v_0 \in L^\infty(\mathbb{R}^n)$; the solution v^ε converges uniformly on compact sets to the viscosity solution v of the homogenized Hamilton–Jacobi equation

$$v_t + \overline{H}(\nabla v) = 0, \quad x \in \mathbb{R}^n \times (0, +\infty), \qquad (2.57)$$

where the homogenized Hamiltonian is defined through solving the cell problem stated below.

Theorem 2.6. *For each $p \in \mathbb{R}^n$, there exits a unique real number $\overline{H}(p)$ such that the equation $H(p + \nabla w, y) = \overline{H}(p)$ has a 1-periodic viscosity solution $w = w(y)$.*

The solution v^ε has the two-scale expansion

$$v^\varepsilon \sim v_0(x, t) + \varepsilon v_1(x, \varepsilon^{-1}x, t) + \cdots, \qquad (2.58)$$

implying to leading order upon substitution in (2.56) that

$$v_{0,t} + H(\nabla_x v_0(x, t) + \nabla_y v_1(x, y, t)) = 0, \qquad (2.59)$$

which leads to the cell problem in Theorem 2.6, where $y = x/\varepsilon$ is the variable, (x, t) are parameters. Equation (2.59) is a nonlinear eigenvalue problem, producing the cell problem in terms of the variable y, and the homogenized equation (2.57) in the variables (x, t).

The ansatz (2.58) is utilized in the convergence proof of [148]. For generalizations to fully nonlinear first- and second-order equations, see [79], where a weak convergence method called the perturbed test function method is employed. Such a method incorporates the above ansatz in the structures of the test functions instead, and can handle equations of first and second order in a unified way.

The homogenized Hamiltonian $\overline{H}$ is convex if H is in p, but it may lose strict convexity. One example [148] is that the homogenized Hamiltonian $\overline{H}$ of the strictly convex classical Hamiltonian $H(p,x) = p^2/2 + V(x)$ is flat near $p = 0$. In fact, let $V \leq 0$ and $\max V = 0$. The cell problem reads

$$\frac{1}{2}(p + w_y)^2 + V(y) = \bar{H}, \quad y \in T^1,$$

which is solvable and gives $\bar{H} \geq 0$ such that

$$\bar{H} = 0 \quad \text{if } |p| \leq \left\langle \sqrt{-2V} \right\rangle,$$

$$|p| = \left\langle \sqrt{2\bar{H} - 2V(y)} \right\rangle \quad \text{if } |p| > \left\langle \sqrt{-2V} \right\rangle, \tag{2.60}$$

where $\langle \cdot \rangle$ denotes the average over one period.

Now we return to equation (2.55) with $c(\varepsilon^{-1}x)$ in place of the last nonlinear term. Using the above homogenization ansatz (2.58), it is straightforward to derive the cell problem

$$\overline{H} = \frac{1}{2} \sum_{i,j=1}^n a_{ij}(y)w_{y_iy_j} + \frac{1}{2} \sum_{i,j=1}^n a_{ij}(y)(p_i + w_{y_i})(p_j + w_{y_j}) + \sum_{i=1}^n b_i(p_i + w_{y_i}) + c(y),$$

$$\tag{2.61}$$

where we solve for a periodic function w and a real constant $\overline{H}$ for given p. The homogenized equation is $v_t - \overline{H}(\nabla v) = 0$. The cell problem (2.61) can be transformed into a linear eigenvalue problem with $\overline{H}$ the principal eigenvalue. To see this, let $\bar{w} = e^w > 0$. Then (2.61) in terms of $\bar{w}$ reads

$$\overline{H}\bar{w} = \frac{1}{2} \sum_{i,j=1}^n a_{ij}\bar{w}_{y_iy_j} + \sum_{i,j=1}^n a_{ij}p_i\bar{w}_{y_j} + \sum_{i,j=1}^n b_i(p_i\bar{w} + \bar{w}_{y_i})$$

$$+ \frac{1}{2} \sum_{i,j=1}^n a_{ij}p_ip_j\bar{w} + c(y)\bar{w}. \tag{2.62}$$

The right-hand-side operator in (2.62) is just L_{-p}, in view of (2.51). Hence $\overline{H}(-z) = \lambda(z)$.

To derive the front speed formula (2.52), consider the Hamilton–Jacobi equation

$$v_t - \overline{H}(\nabla v) = 0$$

with initial condition

$$v_0(x) = \begin{cases} 0 & \text{if } x \in G_0, \\ -\infty & \text{otherwise.} \end{cases}$$

The Hopf formula is

$$v(x,t) = -\inf_{y \in G_0} \overline{H}^*\left(\frac{y-x}{t}\right), \tag{2.63}$$

where $\overline{H}^*$ is the Legendre transform of $\overline{H}$. The function $H(y)$ in the large-deviation approach is related to $\overline{H}^*$ by

$$H(y) = \sup_{-z \in \mathbb{R}^n} [(y, -z) - \lambda(-z)] = \sup_{z \in \mathbb{R}^n} [(-y, z) - \overline{H}(z)] = \overline{H}^*(-y).$$

The points (x, t) where $v < 0$ or $\lim_{\varepsilon \to 0} u^\varepsilon = 0$ then satisfy

$$\overline{H}^* \left(\frac{y - x}{t} \right) > 0, \quad \forall y \in G_0.$$

Since G_0 is compact, we can take both x and t large compared with the size of G_0. Then we drop y to get the condition

$$\overline{H}^* \left(-\frac{x}{t} \right) = H \left(\frac{x}{t} \right) > 0,$$

implying that the front speed $v(e)$ along direction e satisfies $H(v(e)e) = 0$.

Putting the homogenization ansatz (2.58) into (2.54) shows that for KPP fronts in periodic media, the solution u^ε behaves like

$$u^\varepsilon(t, x) = \exp\{-I(t, x, \varepsilon)/\varepsilon\} + \cdots,$$
$$I(t, x, \varepsilon) = I_0(t, x) + \varepsilon I_1(t, x, x/\varepsilon) + \cdots, \qquad (2.64)$$

where I can be regarded as a phase function as in a geometric optics (Wentzel–Kramers–Brillouin (WKB)) ansatz [237]. However, for fronts of type 3 and type 5, the ansatz for u^ε in the same scaling ($x \to \varepsilon^{-1}x$, $t \to \varepsilon^{-1}t$) is

$$u^\varepsilon(t, x) = U(\varphi(t, x, \varepsilon)/\varepsilon, x/\varepsilon) + \cdots,$$
$$\varphi(t, x, \varepsilon) = \varphi_0(t, x) + \varepsilon \varphi_1(t, x) + \cdots, \qquad (2.65)$$

where $\varphi(t, x, \varepsilon)$ is the phase variable. Plugging (2.65) into (2.50), we have

$$\frac{1}{2} (\nabla_x \varphi_0 \partial_s + \nabla_y)(a(y)(\nabla_x \varphi_0 \partial_s + \nabla_y)U) + b(y) \cdot (\nabla_x \varphi_0 \partial_s + \nabla_y)U$$
$$- \varphi_{0,t} U_s + f(U) = 0, \qquad (2.66)$$

where $U = U(s, y)$, $s = \varphi(t, x, \varepsilon)/\varepsilon$, $y = x/\varepsilon$. We see that (2.66) is just the traveling-front equation (2.18) with $k = \nabla_x \varphi_0$, and $c(k) = -\varphi_{0,t}$. Relating them gives the Hamilton–Jacobi equation

$$\varphi_{0,t} + c(\nabla_x \varphi_0) = 0 \qquad (2.67)$$

for the general front evolution. By uniqueness of c in the case of type $(3, 5)$, it is seen from (2.18) that $c = c(k)$ is homogeneous of degree 1 in k, so $c(\nabla_x \varphi_0) = |\nabla_x \varphi_0| c(v_n)$, where $v_n = \nabla_x \varphi_0 / |\nabla_x \varphi_0|$ is the normal direction of the level set of φ_0. The effective Hamiltonian of type-$(3, 5)$ nonlinearities is anisotropic and has linear growth in $|p|$. In contrast, the effective Hamiltonian $\bar{H}$ of KPP in (2.62) is quadratic in p.

Interestingly, there are mechanical analogies of quadratically and linearly growing Hamiltonians. The Hamiltonian of classical mechanics $H = |p|^2/2 + V(x)$ is quadratic in $|p|$. The Lagrangian of a special relativistic particle of mass m in a scalar potential [103, 139] is

$$L(q,x) = -mc^2\sqrt{1 - |q|^2/c^2} - V(x), \quad |q| \in [0,c],$$

where c is the speed of light. The corresponding Hamiltonian is

$$H(p,x) = mc^2\sqrt{1 + |p|^2/c^2} + V(x),$$

which has linear growth in $|p|$. Bistable and ignition-type fronts belong to the family of special relativity, while KPP fronts are Newtonian.

2.5 Fronts in Multiscale Media

The study of traveling fronts in heterogeneous media has been an active area of research in recent years. One may find that other equations also have periodically varying traveling waves (2.16), and one may also consider more complicated media arising in applications, such as space–time-periodic media, time- or space-almost-periodic media, or more general and complex media. We shall present some of these extensions here. An interesting trend is that extended front equations become more and more degenerate if we continue to construct their time dependence explicitly (e.g., constant-speed motion), and that one must use more general dynamic variables to capture these fronts. More complicated media make more complicated fronts.

Recall the solute transport equation (1.1) in the introduction of Chapter 1. In one spatial dimension, v is a constant, and equation (1.1) simplifies after a rescaling of constants to

$$(u + k(x)u^p)_t = (D(x)u_x)_x - u_x, \tag{2.68}$$

where we also make D spatially dependent. We consider the boundary conditions $u(-\infty,t) = u_l$, $u(+\infty,t) = u_r = 0$, $0 < u_l$, representing constant input of solute from the left end of a solute-free soil column. Solutions of (2.68) under such boundary conditions give rise to front solutions.

If k and D are constants, then by making the change of variable $v = u + ku^p$, we can write (2.68) as a standard conservation law:

$$v_t + (f(v) - (g(v))_x)_x = 0, \quad x \in \mathbb{R}. \tag{2.69}$$

Front solutions $v = v(x - ct)$ are solvable in closed form, and $c = (f(u_l) - f(u_r))/(u_l - u_r)$ is the so-called Rankine–Hugoniot relation.

Let us now consider periodic media by supposing $k(x)$ and $D(x)$ to be 1-periodic regular functions. In periodic media, just as in reaction–diffusion equations, traveling fronts take the form $u = U(x - ct, x)$, which turn out to exist also for conservative equations such as (2.68) and are asymptotically stable [244, 245]:

Theorem 2.7. *Let $k(x)$ and $D(x)$ be smooth positive functions with period 1. If $u_r = 0 < u_l$, then equation (2.68) admits a Hölder continuous traveling wave solution of the form $u = U(x - st, x) \equiv U(\xi, y)$, $\xi = x - st$, $y = x$, $U(-\infty, y) = u_l$, $U(+\infty, y) = 0$, and $U(\xi, \cdot)$ has period 1. Such solutions are unique up to constant translations in ξ, and have wave speeds*

$$s = \frac{u_l}{u_l + \langle k \rangle f(u_l)} > 0, \tag{2.70}$$

with $\langle k \rangle$ the periodic mean. The wave profile U satisfies

$$0 \le U < u_l \ \forall(\xi, y); \quad U(\xi_1, y) \le U(\xi_2, y) \ \forall \xi_1 \ge \xi_2, \forall y; \quad U_\xi < 0 \ \text{if} \ U(\xi, y) > 0.$$

Assume that the initial condition $u_0(x)$ satisfies

$$0 \le u_0(x) \le u_l, \ u_0 \in L^1(\mathbb{R}^+); \ u_0^p \in L^1(\mathbb{R}^+), \ u_0 - u_l \in L^1(\mathbb{R}^-), \ u_0^p - u_l^p \in L^1(\mathbb{R}^-).$$

Let also $m(u,x) = u + k(x)u^p$. Then there exists a unique number x_0 such that

$$\int_{\mathbb{R}} m(u_0(x), x) - m(U(x + x_0, x), x) \, dx = 0 \tag{2.71}$$

and such that

$$\lim_{t \to \infty} \|u(t, x) - U(x - st + x_0, x)\|_1 = 0. \tag{2.72}$$

The construction of traveling waves uses the continuation method, and the existence result holds also in several spatial dimensions [244]. The Hölder continuity of solutions is a consequence of u^p being nondifferentiable at $u = 0$. The explicit effective wave speed (2.70) is due to the fact that equation (2.68) is conservative. Only the mean value of k contributes to the speed; the rest of the information in k influences the wave profile.

The stability proof extends that of [188] and uses L^1 contraction of dynamics, as well as a space–time translation invariance of the traveling fronts in the moving frame coordinate. For fronts in another conservative equation (the Richards equation of water infiltration) with more complicated dependence of wave speeds on the periodic media, see [88].

Space–time-dependent media (flows) arise in combustion [10, 65]. KPP fronts in a periodic flow field with space–time-separated scales are studied in [152], which considers the temperature field of a reacting passive scalar,

$$T_t^\varepsilon + V(x, t, \varepsilon^{-\alpha}x, \varepsilon^{-\alpha}t) \cdot \nabla T^\varepsilon = \varepsilon \kappa \Delta T^\varepsilon + \varepsilon^{-1} f(T^\varepsilon), \tag{2.73}$$

with compactly supported (in G_0) nonnegative initial data, and $\alpha \in (0, 1]$. The velocity V is bounded and Lipschitz continuous and has periodic dependence on the fast-oscillating scales $y \equiv \varepsilon^{-\alpha}x$, $\tau \equiv \varepsilon^{-\alpha}t$. The small parameter ε measures the ratio of the front thickness and large scale (dependence on (x,t)) of the velocity field, say of order $O(1)$. The effective Hamiltonian $H(p,x,t)$ is defined as a solution of the following cell problem: for each $(p,x,t) \in \mathbb{R}^n \times \mathbb{R}^n \times (0,+\infty)$ there are a unique number $H(p,x,t)$ and a $w(y,\tau) \in C^{0,1}(\mathbb{R}^n \times (0,+\infty))$ periodic in both y and τ such that

$$w_\tau - a(\alpha)\kappa \Delta w - \kappa |p + \nabla w|^2 + V(x,t,y,\tau) \cdot (p + \nabla w) = -H(p,x,t), \qquad (2.74)$$

where $a(\alpha) = 0$ if $\alpha \in (0,1)$, $a(\alpha = 1) = 1$.

The case $\alpha = 1$ can be derived using an exponential change of variable and a Hamilton–Jacobi equation as in the last section except that due to the time dependence, the w_τ term is added. The condition $a(\alpha) = 0$ in the case $\alpha \in (0,1)$ implies the loss of viscosity in the cell problem (2.74), which can be understood as follows. Ignore the slow variable (x,t) for now and change the scaling to $x = \varepsilon^{-1+\alpha} x'$. Then the velocity V is $V(\varepsilon^{-1}x', \varepsilon^{-1}t')$, and the diffusion coefficient becomes $\varepsilon^{3-2\alpha}\kappa \ll \varepsilon\kappa$. Hence the diffusion term is too small to be seen at the order of the cell problem.

The function H is locally Lipschitz continuous, convex in p, and grows quadratically in $|p|$ as $|p| \to +\infty$ uniformly in (x,t). The asymptotics of T^ε as $\varepsilon \to 0$ are given by the following theorem.

Theorem 2.8. *Let T^ε be a solution of (2.73) under the above assumptions. Then as $\varepsilon \to 0$, $T^\varepsilon \to 0$ locally uniformly in $\{(x,t) : Z < 0\}$ and $T^\varepsilon \to 1$ locally uniformly in the interior of $\{(x,t) : Z = 0\}$, where $Z \in C(\mathbb{R}^n \times [0,+\infty)$ is the unique viscosity solution of the variational inequality*

$$\max(Z_t - H(\nabla Z, x, t) - f'(0), Z) = 0, \quad (x,t) \times \mathbb{R}^n \times (0,+\infty),$$

with initial data $Z(x,0) = 0$ in G_0 and $Z(x,0) = -\infty$ otherwise. The set $\Gamma_t = \partial\{x \in \mathbb{R}^n : Z(x,t) < 0\}$ can be regarded as a front.

Given a space–time-periodic incompressible flow field, the "cell problem" for KPP front speed in the limit $t \to +\infty$ is always viscous. To show this, let us consider

$$u_t = \Delta u + b(x,t) \cdot \nabla u + f(u), \qquad (2.75)$$

where $x \in \mathbb{R}^N$, $t \in \mathbb{R}$, and f is of KPP type. The N components of the vector field $b(x,t) := (b^1(x,t), b^2(x,t), \ldots, b^N(x,t))$ are smooth and spatially divergence-free, are periodic of period 1 in both x and t, and have mean zero over the period cell $Q \times (0,1)$, where Q is the unit cube in $\mathbb{R}^N$. Then the KPP large-time minimal front speed in direction k (denoted by $c^*(k)$) is identified by a front propagation (front spreading) theorem similar to Theorem 2.3. It is given by the variational formula [173]

$$c^*(k) = \inf_{\lambda > 0} \mu(\lambda, k)/\lambda,$$

where $\mu(\lambda, k)$ is the principal eigenvalue of the periodic–parabolic operator [117]

$$L^\lambda \Phi := \Delta_x \Phi + (b - 2\lambda k) \cdot \nabla_x \Phi + \left(\lambda^2 - \lambda b \cdot k + f'(0)\right)\Phi - \Phi_t, \qquad (2.76)$$

defined on spatially–temporally periodic functions $\Phi(x,t)$. The eigenvalue problem (2.76) is related to (2.74) at $\alpha = 1$ by a logarithmic transform. In the case $\alpha = 1$, taking $\varepsilon \to 0$ is the same as $t \to \infty$.

Moreover, there is a traveling-front solution of the form $u = U(k \cdot x - c^*(k)t, x, t) \equiv U(s,x,t)$, locally integrable in (s,x,t), periodic in (x,t), $U(\pm\infty,x,t) = 0/1$, with the continuous directional derivatives

$$U_\tau - c^* U_s, \quad k^i U_s + U_{y^i}, \quad i = 1,\ldots,N, \quad \text{and} \quad (k\partial_s + \nabla_y)^2 U$$

and satisfying the traveling-front equation

$$U_\tau - c^* U_s = (k\partial_s + \nabla_y)^2 U + b \cdot (k\partial_s + \nabla_y) U + f(U). \tag{2.77}$$

Note that (2.77) is an extension of (2.18), and is more degenerate in the sense that there are not enough derivatives in (2.77) to ensure continuity (smoothness) of U. The function U has one more dependent variable than u, which does not happen in spatially periodic media. The value c^* is minimal in that no solutions exist if c^* is replaced by a number $c < c^*$. Recently, it was proved in [168] that for almost all $\eta \in \mathbb{R}$, we have that $U(k \cdot x - c^*(k)t + \eta, x, t)$ satisfies (2.75). Similar existence results [173, 168] hold for type $(2,4,5)$. We refer to [168] for further results on existence of KPP fronts at $c > c^*$ and continuous KPP fronts.

In view of (2.74) and (2.76), the front speed obtained from the limit $t \to \infty$ at a fixed $\varepsilon > 0$ and that from $\varepsilon \downarrow 0$ at a fixed time interval $[0,T]$ may not agree in general. A numerical study of the difference due to finite ε (finite front thickness) is carried out in [161]. At any finite $\varepsilon > 0$, the cell problem of a front is always viscous, while it is not in the limit $\varepsilon \downarrow 0$ when $\alpha < 1$. This shows the subtlety of front speed upscaling in heterogeneous media. The front speed would be the same from either limit in homogeneous media.

Fronts of the form (2.16) persist in space-almost-periodic media [168]. Front solutions of the form $u = U(k \cdot x - ct, t)$ in time-periodic media have been studied in [5] (bistable f) and [97] (nonnegative f). Bistable fronts of the form $u = U(k \cdot x - ct, t)$ have been found in time-almost-periodic media. Interestingly, there are also fronts of the form $u = U(k \cdot x + \int_0^t c(s)\,ds)$ where $c(s)$ is almost periodic. The latter fronts are not reducible to the former. KPP fronts in space-periodic and time-almost-periodic media were studied recently [120], where generalized speed intervals were proved to exist, and they reduce to singletons in the case of time-periodic media. A more general form of fronts has been introduced recently [21, 159, 219], and proved to exist in various settings [219, 163, 174]. For fronts in periodic media in the context of discrete models, see [236, 106] and references therein. For fronts and homogenization in the context of free-boundary limits and models, see [46, 47, 129, 130, 248, 249]. For fronts in periodically perforated media and fragmented environments among other applications, see [113, 25]. For KPP speeds c_* under various parameter asymptotic limits (diffusion–reaction rates, periods), see [225]. For pulselike waves of reaction–diffusion systems in heterogenous media, see [171, 252].

2.6 Variational Principles, Speed Bounds, and Asymptotics

The KPP variational principle (2.52) reduces the front speed problem to the analysis and estimation of the principal eigenvalues of linear advection–diffusion operators, where many classical methods apply. Let us consider (2.75) in space dimension two $(N = 2)$, and scale the velocity field $b(x,t)$ to δb. If δ is small, a perturbation analysis of eigenvalues yields the quadratic enhancement law $c_* = c_0 + O(\delta^2)$, where c_0 is the KPP front speed in homogeneous media. The quadratic correction is explicit in the case of shear flow [191, 177]. In the case of spatial shear, write $b = (0, b_2(x_1))$, $b_2 = \tilde{b}_{x_1}$. The function $\tilde{b}$ has mean equal to zero and serves as the velocity potential. Then the front speed along the x_2 direction has the expansion [191]

$$c_* = c_0 \left(1 + \frac{1}{2} \|\tilde{b}\|_2^2 \delta^2 + \text{higher-order terms} \right), \quad \delta \ll 1. \tag{2.78}$$

The energy (half of L^2 norm square) of $\tilde{b}$ (velocity potential) is the amount of enhancement to leading order. In the case of bistable nonlinearity, one may perform a perturbation analysis of the traveling-front equation (a nonlinear eigenvalue problem) [191]. The interesting finding is that the correction term in (2.78) is the same. By monotonicity of c_* in terms of f, the value of c_* from other types of f must behave the same. This is the first indication that for front speed c_*, the type of nonlinearity does not matter as much as the flow. In other words, there is *universality of c_* in terms of nonlinearity*.

For time-periodic shear flow, let

$$b_2 = b_2(x_1, t) = \sum_{m \neq 0, l \neq 0} b_{m,l} e^{imx_1 + i\omega lt}.$$

Then

$$c_* = c_0 \left(1 + \frac{1}{2} \left(\sum_{\substack{m>0 \\ l>0}} |b_{m,l}|^2 \frac{2m^2}{m^4 + l^2 \omega^2} \right) \delta^2 + \text{higher-order terms} \right). \tag{2.79}$$

We see that the enhancement decreases with increasing frequency of temporal oscillations (or as ω increases). The speed slowdown due to temporal oscillations is called the speed-bending phenomenon in the combustion literature, and is studied in various models [10, 128, 65]. It persists in random flows as well [65, 179, 182], which we shall discuss more in Chapter 5. Again formula (2.79) holds for all nonlinearities.

In the large $\delta \gg 1$ regime, consider again spatial shear flow. The eigenvalue analysis of KPP front speeds [19] shows that $c_*(\delta)/\delta$ is monotone decreasing in $\delta \gg 1$ and converges to a positive limit. The limiting value or the linear growth rate depends on $\tilde{b}$ in an implicit way, and it has a variational formula [115]:

$$\lim_{\delta \to \infty} c_*(\delta)/\delta = \sup_{\psi \in D_1} \int_\Omega \tilde{b}(x_1)\, \psi^2(x_1)\, dx_1, \qquad (2.80)$$

where

$$D_1 = \left\{ \psi \in H^1(\Omega) : \|\nabla \psi\|_2^2 \le f'(0), \|\psi\|_2 = 1 \right\},$$

and Ω is the periodic domain of variable x_1. If $\tilde{b}$ has a flat piece near its maximal point in Ω, a test function in D_1 can be supported near the maximal point, and the limit equals $\max_\Omega \tilde{b} = \|\tilde{b}\|_\infty$. If the reaction is fast ($f(u)$ replaced by $rf(u)$, $r \gg 1$) or the diffusion constant (equal to one in (2.75)) is made small, the constraint $\|\nabla \psi\|_2^2 \le f'(0)$ is easy to satisfy: again a test function may be localized near the maximal point of $\tilde{b}$, and so the speed growth rate is close to $\max_\Omega \tilde{b}$ [12, 24]. In general,

$$c_* = O(\delta), \qquad \delta \gg 1, \qquad (2.81)$$

for other nonnegative nonlinearities. Here again we see universal behavior. The linear law (2.81) holds also for time-periodic shear flows [177], and is numerically observed for bistable f as well. It is true for more general percolating flows that contain at least two infinitely long channels of flow trajectories [59, 132]. The open channels (streamlines) in the flow are like multiple lanes on the freeway to help the transport process and speed up the reaction front. The growth exponent is less than one (sublinear growth) for flows with enough closed streamlines. For example,

$$c_*(\delta) = O(\delta^{1/4}), \qquad \delta \gg 1, \qquad (2.82)$$

holds [12, 184] for the KPP front and cellular flow:

$$b = (-\phi_{x_2}, \phi_{x_1}), \qquad \phi = \cos(\pi x_1)\cos(\pi x_2). \qquad (2.83)$$

The proof of the $\frac{1}{4}$ scaling for the KPP front [184] uses the speed variational formula (2.52), boundary layer analysis of cellular flows, and properties of convection-enhanced diffusion [86]. For ignition nonlinearity, (2.82) is supported by numerical simulations [231]. Moreover, analytical bounds $O(\delta^{1/5}) \le c_* \le O(\delta^{1/4})$ hold [132]. A useful criterion for distinguishing linear and sublinear speed growth is in terms of first integrals of the flow fields [24]. A first integral for a periodic vector field b is a nonzero periodic solution w of the equation $b \cdot \nabla w = 0$. An extension of (2.80) to KPP speed c_* in a mean-zero divergence-free vector field $b(x)$ is [256]

$$\lim_{\delta \to \infty} c_*(\delta, k)/\delta = \sup_{w \in D_I} \int_{T^N} (b \cdot k)\, w^2(x)\, dx,$$

where

$$D_I = \left\{ w \in H^1(T^N) : b \cdot \nabla w = 0, \|\nabla w\|_2^2 \le f'(0), \|w\|_2 = 1 \right\}, \qquad (2.84)$$

where k is the direction of front propagation and T^N the N-dimensional unit torus (a unit cube in $\mathbb{R}^N$ with opposite faces identified).

It follows that an upper bound is $\|b \cdot k\|_\infty$, and that the limit in (2.84) is nonzero if $\int_{T^N} (b \cdot e) w_0 \, dx \neq 0$ for some first integral w_0. This is the case for shear flows.

In general, suppose such a w_0 exists. Then $w = (1 + \varepsilon w_0)/\|1 + \varepsilon w_0\|_2 \in D_I$ for ε small enough, and the limiting value is to leading order $\varepsilon \int_{T^N} (b \cdot e) w_0 \, dx$, which is positive if ε is chosen to have the sign of the integral. If $\int_{T^N} (b \cdot k) w^2 \, dx \leq 0$ for all first integrals, then $c_*(\delta, k) = o(\delta)$. The cellular flow (2.83) is an example for which $\int_{T^N} (b \cdot k) w^2 \, dx = 0$ for all k and first integral w.

The front asymptotic enhancement in the sense of $\lim_{\delta \to \infty} c_*(\delta) = \infty$ by periodic incompressible flow has been shown recently [255] to depend on the geometry of the flow and not on nonlinearity f. In particular, front asymptotic enhancement occurs for KPP if and only if it does so for ignition f, and so the phenomenon is universal among all nonnegative reactions.

A variant of (2.52) holds for partially periodic media where solutions in part of the variables are periodic and in the other part are subject to zero Neumann boundary conditions [20]. There are min–max variational principles of front speeds [109, 116, 232] for non-KPP f. In particular, let us state the one for the unique front speed from shear flow and bistable/ignition f that has been used in analysis of random front speeds [176].

Consider the cylindrical domain $x = (x_1, \tilde{x}) \in D = \mathbb{R} \times \Omega$, where Ω is a bounded domain in $\mathbb{R}^{N-1}$, and shear flow $b = (b_1(\tilde{x}), 0)$. The front moves along x_1, $u = U(x_1 + c_* t, \tilde{x})$, satisfying zero Neumann boundary condition at $\mathbb{R}^1 \times \partial \Omega$. The initial datum u_0 belongs to the set I_s. The set I_s for bistable f consists of bounded continuous functions with limits one and zero at $x_1 \sim \pm \infty$ respectively. For ignition f, one requires also that u_0 decay to zero exponentially at $-\infty$. For $u_0 \in I_s$, $u(x, t)$ converges to a traveling front at large times [205]. Define the functional as in [116]:

$$\psi(v) = \psi(v(x)) \equiv \frac{Lv + f(v)}{\partial_{x_1} v} \equiv \frac{\Delta v + b_1(\tilde{x}) \partial_{x_1} v + f(v)}{\partial_{x_1} v}. \tag{2.85}$$

The min–max variational formula [116] for c_* is

$$\sup_{v \in K} \inf_{x \in D} \psi(v(x)) = c(\delta) = \inf_{v \in K} \sup_{x \in D} \psi(v(x)), \tag{2.86}$$

where K is the set of admissible functions,

$$K = \left\{ v \in C^2(D) \mid \partial_{x_1} v > 0, 0 < v(x) < 1, v \in I_s \right\}.$$

The proof uses asymptotic stability of traveling fronts and min–max front speed formulations of [232]. Likewise, similar min–max formulas have been derived and studied for the homogenized Hamiltonian $\bar{H}$ of HJ in periodic media [57, 104]. In [104], $\bar{H}$ is computed based on the formula

$$\bar{H}(p) = \inf_{\phi(y) \in C^1(T^N)} \sup_y H(p + \nabla_y \phi(y), y). \tag{2.87}$$

2.7 Exercises

1. Show that the principal eigenvalue $\lambda(z)$ of the operator L_z in (2.51) is positive for all $z \in \mathbb{R}^n$ if (a_{ij}) is the identity matrix and $b_j(y)$ is a mean-zero and divergence-free vector field.

2. Prove by the maximum principle that the homogenized Hamiltonian of KPP nonlinearity $\bar{H} = \bar{H}(p)$ defined in (2.62) grows like $O(|p|^2)$ for large $|p|$.

3. Show that for cellular flow (2.83), the integral $\int_{T^N} (b \cdot k) w^2 \, dx$ is zero for all unit vectors $k \in \mathbb{R}^2$ and first integral w ($b \cdot \nabla w = 0$).

4. Derive the quadratic speed-enhancement formula (2.78) for bistable and ignition fronts with the min–max formula (2.86) in a mean-zero 1-periodic shear flow $b = \delta(b_1(\tilde{x}), 0)$. The fronts move in the x_1 direction, $\tilde{x} \in \mathbb{R}^{n-1}$, $n \geq 2$. In the small-δ regime, define the test function as a perturbation of the traveling-front profile in homogeneous media of the form

$$v(x) = U(\xi) + \delta^2 w(\xi, \tilde{x}), \tag{2.88}$$

where

$$\xi = (1 + \alpha \delta^2) x_1 + \delta \chi, \tag{2.89}$$

with α a constant to be determined and $\chi = \chi(\tilde{x})$ the mean-zero periodic solution of

$$-\Delta_{\tilde{x}} \chi = b_1.$$

Choose α properly so that w is uniformly bounded with decay at $x_1 \sim \pm \infty$ (and so v is admissible) and gives the quadratic speed correction in (2.78).

Chapter 3
Fronts in Random Burgers Equations

In groundwater and contaminant transport problems arising from environmental sciences, fronts typically travel in a spatially inhomogeneous environment because of the natural formation of porous structures. Due to lack of field data, the spatially inhomogeneous environment is often modeled as a random process. Conservation of mass then leads to a nonlinear scalar conservation law with a random flux,

$$U_t + (f(U, x, \omega))_x = 0, \tag{3.1}$$

or its viscous analogue. Some equations of this form are (1) the Buckley–Leverett equation for two-phase flows [118] and references therein; (2) the contaminant transport equation [253, 38]; and (3) the Richards equation for infiltration problems [195, 196, 88]. The specific form of the nonlinear and random function f depends on the problem at hand.

One of the fundamental issues discussed in these works is front dynamics in random media. In this chapter, we shall analyze a special case in which f is quadratic in U with random multiplicative coefficient, or a Burgers equation with random flux. We find that at large times, front motion obeys the central limit theorem. In other words, a front moves at an average deterministic velocity and the fluctuations around it obey Gaussian statistics.

3.1 Main Assumptions and Results

We are interested in the long-time behavior of the inviscid Burgers equation with a random flux,

$$U_t + \left(\frac{1}{2} a(x, \omega) U^2 \right)_x = 0, \tag{3.2}$$

and for initial data of the front type,

$$U(x, 0) = I_{\mathbb{R}_-}(x). \tag{3.3}$$

J. Xin, *An Introduction to Fronts in Random Media*, Surveys and Tutorials in the Applied Mathematical Sciences 5, DOI: 10.1007/978-0-387-87683-2_3,
© Springer Science + Business Media, LLC 2009

Here $I_{\mathbb{R}_-}$ denotes the indicator function of the negative real line, and $a(x, \omega)$ is a stochastic process on the real line satisfying the assumptions A1–A5 stated below.

A1. Stationarity: the finite-dimensional distributions of the process $a(x, \omega)$ are invariant under translations of the variable x.

A2. Positivity: $a(x, \omega) > 0$ with probability one.

A3. Measurability and integrability of the inverse: paths of a are measurable functions of x and $E\left[1/a(x)\right] < \infty$. It follows that also

$$E\left[\frac{1}{\sqrt{a(x)}}\right] \overset{\text{def}}{=} \mu < \infty.$$

A4. Invariance principle: Let

$$\xi(x) = \int_0^x \frac{1}{\sqrt{a(y)}}\, dy.$$

Note that $\xi(x) < 0$ for $x < 0$. For each $x_0 > 0$, we have

$$\left(\frac{\xi(tx) - \mu tx}{\sigma\sqrt{t}}\right)_{|x| \le x_0} \overset{\mathrm{d}}{\to} (W_x)_{|x| \le x_0}, \tag{3.4}$$

as $t \to \infty$, where $W = (W_x)_{x \in \mathbb{R}}$ is the Wiener process and

$$\sigma^2 = 2 \int_0^{+\infty} E\left[\left(\frac{1}{\sqrt{a(0)}} - \mu\right)\left(\frac{1}{\sqrt{a(x)}} - \mu\right)\right] dx < \infty,$$

and $\overset{\mathrm{d}}{\to}$ denotes convergence of processes in law [36]. The finiteness of the last integral is part of the assumption, and σ^2 is sometimes called the velocity autocorrelation function (of the process $1/\sqrt{a}$).

A5. Regularity: the paths of the process a are Hölder continuous with some (positive) exponent. This will be used in the proof of the main theorem to justify taking the zero-viscosity limit. A well-known probabilistic condition that implies Hölder continuity of sample paths is the Kolmogorov moment condition [202, Theorem 25.2].

A large class of processes for which (3.4) holds is the class of stationary processes $a(x, \omega)$ satisfying the appropriate ϕ-mixing condition. Here ϕ is a nonnegative function of a positive real variable such that

$$\lim_{t \to +\infty} \phi(t) = 0, \tag{3.5}$$

and the ϕ-mixing condition says that for any $t > 0$ and for any s, whenever an event E_1 is in the σ-field generated by the random variables $a(x)$ with $-\infty \leq x \leq s$ and an event E_2 is in the σ-field generated by $a(x)$ with $s + t \leq x \leq +\infty$, we have

$$|P[E_1 \cap E_2] - P[E_1]P[E_2]| \leq \phi(t)P[E_1]. \tag{3.6}$$

Roughly speaking, because of (3.5), (3.6) expresses a decay of correlations of the variables $a(x)$. More information on ϕ-mixing processes can be found in [36], where it is proved in particular [36, pp. 178–179] that the invariance principle (assumption A4) holds if $\int_0^{+\infty} \sqrt{\phi(t)}\, dt < \infty$.

The Burgers fronts are asymptotically stable for spatially decaying initial perturbations [121]. The following main result of the chapter shows that the front structure is also present in the presence of a random flux. Similar results hold for random initial perturbations [233]. Throughout the chapter, the symbol $\xrightarrow{d}$ denotes convergence in distribution.

Theorem 3.1 (Random Burgers Fronts). *Let $2c = E[a^{-1/2}]^{-2}$ denote the square root-harmonic mean of the process $a(x, \omega)$. Then as $t \to \infty$,*

$$U(\alpha t, t) \xrightarrow{d} 0, \quad \alpha > c, \tag{3.7}$$

$$\sqrt{a(\alpha t)}\, U(\alpha t, t) \xrightarrow{d} \sqrt{2c}, \quad \alpha < c, \tag{3.8}$$

$$\sqrt{a(ct + z\sqrt{t})}U(ct + z\sqrt{t}, t) \xrightarrow{d} X, \tag{3.9}$$

where X is a random variable equal to $\sqrt{2c}$ with probability $\mathcal{N}\left(\frac{\mu^2}{\sigma}z\right)$ and equal to 0 with probability $1 - \mathcal{N}\left(\frac{\mu^2}{\sigma}z\right)$, where $\mathcal{N}(s) = \frac{1}{\sqrt{2\pi}}\int_{-\infty}^{s}\exp\{-\eta^2/2\}\,d\eta$ is the error function.

The first two parts of the theorem say that to leading order, the front speed in the presence of randomness equals c, or a law of large numbers. The last part of the theorem says that the front fluctuations around the speed c obey Gaussian statistics, or a central limit theorem. If we view a Burgers front location as a time-dependent random variable $X(t, \omega)$, then it satisfies the fundamental laws of probability, similar to sums of iid random variables [72].

In the proof of Burgers' front theorem, we will make use of a regularized equation:

$$U_t + \left(\frac{1}{2}a(x)\,U^2\right)_x = \nu\left(\sqrt{a(x)}\left(\sqrt{a(x)}U\right)_x\right)_x, \tag{3.10}$$

where $\nu > 0$ is a positive parameter that we shall send to zero eventually. It is convenient to rewrite this equation in terms of the function

$$u = \sqrt{a(x)}U. \tag{3.11}$$

The equation for u becomes

$$\frac{u_t}{\sqrt{a(x)}} + (u^2/2)_x = v\left(\sqrt{a(x)}u_x\right)_x. \tag{3.12}$$

To simplify the last equation, we change the space variable:

$$\xi = \int_0^x \frac{1}{\sqrt{a(x')}}\,dx'. \tag{3.13}$$

Since this change of variables depends on the realization of the process a, we obtain in this way a stochastic process $\xi(x,\omega)$, which has already been used to state the assumption A4.

The equation for u in the variables (ξ,t) becomes the standard viscous Burgers equation

$$u_t + \left(u^2/2\right)_\xi = vu_{\xi\xi}, \tag{3.14}$$

with the new initial condition

$$u(\xi,0) = \sqrt{a(x(\xi))}I_{\mathbb{R}_-}(\xi). \tag{3.15}$$

It is known that the speed of a (shock) front of the Burgers equation is equal to its height divided by two, the Rankine–Hugoniot condition [141]. This then leads to an intuitive explanation. The asymptotic speed of the front arising from our random initial condition equals one-half of its average height. Calculated in the ξ variable, the speed is

$$\frac{1}{2}\lim_{L\to\infty}\frac{1}{L}\int_{-L}^0 \sqrt{a(x(\xi))}\,d\xi,$$

which, after changing the variable of integration to x, gives

$$\frac{1}{2}\lim_{L\to\infty}\frac{-x(-L)}{L} = \frac{1}{2}E[a^{-1/2}]^{-1}.$$

To recover the front speed in the x variable, we divide this value by $E[a^{-1/2}]$ in view of (3.13) and arrive at the speed c in the theorem. A similar but more detailed argument, taking into account fluctuations of the total mass in a finite interval of the initial data, leads to a heuristic justification of the Gaussian statistics of the front location.

3.2 Hopf–Cole Solutions

The Hopf–Cole formula [237] for u reads

$$u(\xi,t) = \frac{\int_{-\infty}^{+\infty}\frac{\xi-\eta}{t}\exp\left[-\frac{G(\eta,\xi,t)}{2v}\right]d\eta}{\int_{-\infty}^{+\infty}\exp\left[-\frac{G(\eta,\xi,t)}{2v}\right]d\eta}, \tag{3.16}$$

where

$$G(\eta,\xi,t) = \int_0^\eta u(\eta',0)\,d\eta' + \frac{(\xi-\eta)^2}{2t} = x(\eta)I_{\mathbb{R}_-}(\eta) + \frac{(\xi-\eta)^2}{2t}$$

$$= \left(\frac{\eta}{E\left[a^{-1/2}\right]} + \hat{x}(\eta) \right) I_{\mathbb{R}_-}(\eta) + \frac{(\xi-\eta)^2}{2t}. \qquad (3.17)$$

The second equality follows by changing the variable to $x = x(\eta')$, as in (3.13), where x denotes the inverse of ξ. We used the fact that the derivative of ξ is $1/\sqrt{a}$.
Define $u_l = 1/E[a^{-1/2}]$. Then clearly,

$$u_l = \mu^{-1} = \sqrt{2c}. \qquad (3.18)$$

It is convenient to use the suggestive notation u_l (the "left state of u") below.
    The numerator of (3.16) is equal to

$$\int_{-\infty}^0 \frac{\xi-\eta}{t} \exp\left[\frac{-u_l\eta - \hat{x}(\eta) - (2t)^{-1}(\xi-\eta)^2}{2v} \right] d\eta$$

$$+ \int_0^\infty \frac{\xi-\eta}{t} \exp\left[-(2t)^{-1}\frac{(\xi-\eta)^2}{2v} \right] d\eta, \qquad (3.19)$$

which, with the substitution $y = \xi - \eta$, becomes

$$\int_\xi^\infty \frac{y}{t} \exp\left[\frac{-(\xi-y)u_l - (2t)^{-1}y^2 - \hat{x}(\xi-y)}{2v} \right] dy + \int_{-\infty}^{\xi/\sqrt{t}} \eta e^{-\eta^2/4v}\,d\eta$$

$$= \frac{1}{t} \int_\xi^\infty y\exp\left[\frac{-(\xi-\frac{u_l}{2}t)u_l - (2t)^{-1}(y-u_lt)^2 - \hat{x}(\xi-y)}{2v} \right] dy \qquad (3.20)$$

$$+ \int_{-\infty}^{\xi/\sqrt{t}} \eta e^{-\eta^2/4v}\,d\eta.$$

Changing variables $x' = y - u_l t$, the numerator becomes

$$\frac{1}{t} \int_{\xi-u_lt}^\infty (x' + u_lt)\exp\left[\frac{-(\xi-\frac{u_l}{2}t)u_l - (2t)^{-1}x'^2 - \hat{x}(\xi-x'-u_lt)}{2v} \right] dx'$$

$$+ \int_{-\infty}^{\xi/\sqrt{t}} \eta e^{-\eta^2/4v}\,d\eta$$

$$= u_l\exp\left[-\left(\xi-\frac{u_l}{2}t\right)\frac{u_l}{2v} \right] \int_{\xi-u_lt}^\infty \exp\left[-(2t)^{-1}\frac{x'^2}{2v} - \frac{\hat{x}(\xi-x'-u_lt)}{2v} \right] dx'$$

$$+ t^{-1}\exp\left[-\left(\xi-\frac{u_l}{2}t\right)\frac{u_l}{2v} \right] \int_{\xi-u_lt}^\infty x'\exp\left[-(2t)^{-1}\frac{x'^2}{2v} - \frac{\hat{x}(\xi-x'-u_lt)}{2v} \right] dx'$$

$$+ \int_{-\infty}^{\xi/\sqrt{t}} \eta e^{-\eta^2/4v}\,d\eta.$$

Finally, let us introduce a new variable $\eta = \frac{x'}{\sqrt{t}}$ and rearrange the order of the terms to get

$$\int_{-\infty}^{\frac{\xi}{\sqrt{t}}} \eta e^{-\frac{\eta^2}{4\nu}} d\eta + \sqrt{t} u_l e^{-\frac{u_l}{2\nu}(\xi - \frac{u_l}{2}t)} \int_{\frac{\xi - u_l t}{\sqrt{t}}}^{\infty} e^{-\frac{\eta^2}{4\nu} - \frac{\hat{x}(\xi - \sqrt{t}\eta - u_l t)}{2\nu}} d\eta$$

$$+ e^{-\frac{u_l}{2\nu}(\xi - \frac{u_l}{2}t)} \int_{\frac{\xi - u_l t}{\sqrt{t}}}^{\infty} \eta e^{-\frac{\eta^2}{4\nu} - \frac{\hat{x}(\xi - \sqrt{t}\eta - u_l t)}{2\nu}} d\eta$$

$$\equiv A_t + B_t + C_t. \tag{3.21}$$

Likewise, the integral in the denominator equals $\frac{B_t}{u_l} + D_t$, where B_t is as above and

$$D_t = \sqrt{t} \int_{-\infty}^{\xi/\sqrt{t}} e^{-\eta^2/4\nu} d\eta. \tag{3.22}$$

The Hopf–Cole solution formula is put in the form

$$u = \frac{A_t + B_t + C_t}{B_t/u_l + D_t}, \tag{3.23}$$

where B_t, D_t are positive.

3.3 Asymptotic and Probabilistic Preliminaries

Let us first state a Laplace-type asymptotic lemma:

Lemma 3.2. *Let $\varphi_\lambda(u) \in C(\mathbb{R}^1)$, $\varphi_\lambda(u) \to \varphi(u)$, uniformly on compact sets of u as $\lambda \to \infty$, and $C_1 u^2 \le |\varphi_\lambda(u)| \le C_2 u^2$ for some positive constants C_i, $i = 1, 2$, uniformly in $\lambda \to \infty$. The limiting function $\varphi(u)$ is in $C(\mathbb{R}^1)$, $\varphi(u) < \varphi(u_*)$, $\forall u \ne u_*$. Here c_0 and C are positive constants. Then for the probability measures μ_λ with densities*

$$\frac{\exp\{\lambda \varphi_\lambda(u)\} du}{\int_{\mathbb{R}^1} \exp\{\lambda \varphi_\lambda(u)\} du},$$

we have as $\lambda \to +\infty$,

1. *$\mu_\lambda \xrightarrow{d} \delta(u_*)$, the unit mass at u_*;*
2. *the expected value $E_{\mu_\lambda}(u)$ approaches u_*;*
3. *$\lambda^{-1} \log \int_{\mathbb{R}^1} \exp\{\lambda \varphi_\lambda(u)\} du \to \varphi(u_*)$.*

The proof is left as an exercise with hints.

Let us recall next a probabilistic fact that the stochastic process $W_y - y^2/2$, where W_y is the standard Wiener process on the line, attains a unique maximum almost surely [233]. Such a process is also known as Brownian motion with a parabolic shift; see [105] for a detailed study.

We shall make use of a consequence of assumption A4. Note that the paths of the process $\xi(x, \omega)$ are (with probability one) continuous, strictly increasing functions of x. Therefore, each has a continuous inverse, defining another process $x(\xi, \omega)$. Assumption A4 says that the process $\xi(x, \omega)$ satisfies an invariance principle, which in fact implies that the same is also true about the process $x(\xi)$. More precisely,

$$\left(\frac{x(t\xi) - \frac{t\xi}{\mu}}{\mu^{-\frac{3}{2}}\sigma\sqrt{t}} \right)_{|\xi| \le \xi_0} \xrightarrow{\text{d}} (W_\xi)_{|\xi| \le \xi_0}. \tag{3.24}$$

The complete proof can be found at [234, Theorem 4.1]. In the sequel, we will use the following notation for the process $x(\xi)$ with its mean subtracted:

$$\hat{x}(\xi) = x(\xi) - \frac{\xi}{\mu}. \tag{3.25}$$

3.4 Asymptotic Reductions

In the next two propositions we prove that a part of the expression for u goes to zero at large times, thereby asymptotically simplifying the solutions. These propositions will be used in the proof of the theorem, where it will be important that the convergence take place uniformly in v, in the appropriate sense defined below in (3.26). With this in mind, we adopt the following convention about constants: constants independent of v, but depending on the random parameter ω (i.e., on the realization of the random flux), will be denoted by $C(\omega)$, or simply by C. Constants independent of both v and ω will be denoted by c. The actual value of C or c may vary from one line to another.

Proposition 3.3. *We have*

$$\limsup_{t \to \infty} \frac{A_t}{B_t/u_l + D_t} \stackrel{\text{d}}{=} 0.$$

Moreover, convergence is uniform in v in the sense that for every $\varepsilon > 0$, as $t \to \infty$,

$$P\left[\sup_{v \le v_0} \left| \frac{A_t}{B_t/u_l + D_t} \right| > \varepsilon \right] \to 0, \tag{3.26}$$

for any $v_0 > 0$. Note that convergence in distribution to 0 is equivalent to convergence in probability to 0.

Proof. We have for positive ξ,

$$|A_t| \le v \int_{-\infty}^{0} \eta \exp\{-\eta^2/4v\} \, d\eta = c_1 v,$$

and

$$D_t \geq \sqrt{t} \int_{-\infty}^{0} \exp\{-\eta^2/4v\} \, d\eta = \sqrt{v t}\, c_2,$$

so

$$\frac{|A_t|}{D_t} \leq c\sqrt{v} t^{-1/2}, \tag{3.27}$$

with an absolute constant c. For negative ξ, we restrict the integration in the definition of B_t to the interval $0 \leq \eta \leq 1$ and note that since $\hat{x}(u)/\sqrt{|u|}$ converges in distribution to a normal random variable, with probability one there exists an (ω-dependent) constant C such that for all u,

$$\hat{x}(u) \leq C|u|^{2/3}. \tag{3.28}$$

Therefore the integral

$$\int_{\frac{\xi - u_l t}{\sqrt{t}}}^{\infty} e^{-\frac{\eta^2}{4v} - \frac{\hat{x}(\xi - \sqrt{t}\eta - u_l t)}{2v}} \, d\eta \tag{3.29}$$

can be bounded below by

$$e^{-1/4v} \int_{0}^{1} \exp\left(-\frac{\hat{x}(\xi - \sqrt{t}\eta - u_l t)}{2v}\right) d\eta \geq e^{-1/4v} e^{\frac{-C}{v}|\xi - \sqrt{t} - u_l t|^{2/3}}, \tag{3.30}$$

for all t and ξ. This implies that for almost all ω and $t \geq 1$ (uniformly in $\xi \leq 0$), we have

$$B_t \geq \sqrt{t} u_l \exp\left\{-\frac{u_l}{2v}\left(\xi - \frac{u_l}{2}t\right)\right\} e^{-\frac{1}{4v}} e^{-\frac{C}{v}(\xi - \sqrt{t} - u_l t)^{2/3}}$$

$$\geq u_l e^{-\frac{u_l}{2v}(\xi - \frac{u_l}{2}t) - \frac{C}{v}|\xi - t - u_l t|^{2/3}} e^{-1/4v}.$$

The last expression clearly goes to ∞ uniformly in $\xi \leq 0$ as $t \to +\infty$. Since $|A_t|$ is bounded from above by an absolute constant, it follows that

$$\sup_{\xi \leq 0} \frac{|A_t|}{B_t} \to 0, \tag{3.31}$$

for almost all ω. Combining (3.27) and (3.31) ends the proof.   □

Let us now proceed with our next proposition.

Proposition 3.4. *We have*

$$\limsup_{t \to \infty} \sup_{\xi} \frac{C_t}{B_t/u_l + D_t} = 0;$$

the convergence is uniform in $v \in (0, v_0)$ *in the sense of* (3.26).

Proof (Sketch). We begin by observing that

$$\frac{C_t}{B_t} = \frac{1}{\sqrt{tu_l}} \cdot \frac{e^{\frac{-\hat{x}(\xi-u_lt)}{2\nu}} \int_{\frac{\xi-u_lt}{\sqrt{t}}}^{\infty} \eta e^{\frac{[\hat{x}(\xi-u_lt)-\hat{x}(\xi-u_lt-\sqrt{t}\eta)]}{2\nu} - \frac{\eta^2}{4\nu}} \, d\eta}{e^{\frac{-\hat{x}(\xi-u_lt)}{2\nu}} \int_{\frac{\xi-u_lt}{\sqrt{t}}}^{\infty} e^{\frac{[\hat{x}(\xi-u_lt)-\hat{x}(\xi-u_lt-\sqrt{t}\eta)]}{2\nu} - \frac{\eta^2}{4\nu}} \, d\eta}. \tag{3.32}$$

Changing the variable to $y = t^{-\frac{1}{6}}\eta$, we obtain

$$\frac{C_t}{B_t} = \frac{1}{\sqrt{tu_l}} \cdot \frac{t^{1/3} \int_{\frac{\xi-u_lt}{t^{2/3}}}^{\infty} ye^{\frac{t^{1/3}}{2\nu}\left[\frac{\hat{x}(\xi-u_lt)-\hat{x}(\xi-u_lt-t^{2/3}y)}{t^{1/3}} - \frac{y^2}{2}\right]}}{t^{1/6} \int_{\frac{\xi-u_lt}{t^{2/3}}}^{\infty} e^{\frac{t^{1/3}}{2\nu}\left[\frac{\hat{x}(\xi-u_lt)-\hat{x}(\xi-u_lt-t^{2/3}y)}{t^{1/3}} - \frac{y^2}{2}\right]}}. \tag{3.33}$$

We shall first consider the values of ξ satisfying

$$\xi - \frac{2}{3}u_l t \leq 0. \tag{3.34}$$

The stationarity of a implies easily that

$$\frac{\hat{x}(\xi - u_l t) - \hat{x}(\xi - u_l t - t^{2/3}y)}{t^{1/3}} \overset{\mathrm{d}}{=} \frac{\hat{x}(t^{2/3}y)}{t^{1/3}}, \tag{3.35}$$

with equality in law of processes in the variable $y \in \mathbb{R}$. As remarked above, assumption A4 implies that the processes

$$\frac{\hat{x}(\xi - u_l t) - \hat{x}(\xi - u_l t - t^{2/3}y)}{\sigma' t^{1/3}}, \tag{3.36}$$

where $\sigma' = \sigma'(\mu, \sigma)$ is a positive constant, converge in law to the Wiener process, which can be strengthened to almost sure uniform convergence on compact intervals of y by a change of probability space via the Skorohod representation theorem (see [202, Theorem 86.1]). With the Laplace asymptotic lemma (Lemma 3.2), we see that as long as

$$\frac{\xi - u_l t}{t^{2/3}} \to -\infty'$$

the ratio of the two integrals of (3.33) converges in distribution to y_0, where y_0 denotes the unique value of y where the function $\sigma' W_y - y^2/2$ attains its maximum. Existence and uniqueness of such a point implies that for almost all ω, C_t/B_t converges to zero uniformly in ξ satisfying (3.34).

To handle the values of ξ for which

$$\xi - \frac{2}{3}u_l t > 0, \tag{3.37}$$

note that $\xi - \frac{u_l}{2}t \to +\infty$, uniformly in ξ satisfying (3.37). We will now use an argument similar to the one used in the proof of Proposition 3.3 to show that for almost all ω, the quantity C_t/B_t converges to zero uniformly in ξ satisfying (3.37). With probability one, there exists an (ω-dependent) constant C such that (3.28) holds. This, together with subadditivity of the function $u \mapsto |u|^{2/3}$, implies that

$$\hat{x}(\xi - \sqrt{t}\eta - u_l t) \geq -C\left(|\xi|^{2/3} + t^{1/3}|\eta|^{2/3} + u_l^{2/3}t^{2/3}\right).$$

Therefore the ξ-dependent part of the integrand can be absorbed into the prefactor:

$$|C_t| \leq e^{-\frac{u_l}{2\nu}(\frac{5}{6}\xi - \frac{u_l}{2}t)} \int_{\frac{\xi - u_l t}{\sqrt{t}}}^{\infty} \eta e^{-\frac{\eta^2}{4\nu} + \frac{C}{2\nu}(t^{1/3}|\eta|^{2/3} + u_l^{2/3}t^{2/3})} \, d\eta. \tag{3.38}$$

We now divide the integral in the last formula into two parts, corresponding to $|\eta| \leq 1$ and $|\eta| \geq 1$. The first integral is clearly bounded above by

$$ce^{-\frac{u_l}{2\nu}(\frac{5}{6}\xi - \frac{u_l}{2}t)}e^{ct^{2/3}}.$$

The last expression goes exponentially fast to zero, uniformly in ξ satisfying (3.37), since for those ξ,

$$\frac{5}{6}\xi - \frac{u_l}{2}t \geq \frac{1}{18}u_l t. \tag{3.39}$$

When $|\eta| \geq 1$, we have $|\eta|^{2/3} \leq |\eta|$, and consequently, the right-hand side of (3.38) is bounded above by

$$e^{-\frac{u_l}{2\nu}(\frac{5}{6}\xi - \frac{u_l}{2}t)}e^{\frac{C}{2\nu}u_l^{2/3}t^{2/3}}\int_{\frac{\xi - u_l t}{\sqrt{t}}}^{\infty}|\eta|e^{-\frac{\eta^2}{4\nu} + \frac{C}{2\nu}t^{1/3}|\eta|}\,d\eta.$$

The integral in the above formula can be estimated by first absorbing the factor $|\eta|$ into the exponential factor (by making C bigger) and then using the identity

$$\int_{\mathbb{R}} e^{-\frac{\eta^2}{4\nu} + \frac{C}{2\nu}s\eta}\,d\eta = \sqrt{4\pi\nu}e^{Cs}$$

with $s = t^{1/3}$. We obtain in this way

$$|C_t| \leq e^{-\frac{u_l}{2\nu}(\frac{5}{6}\xi - \frac{u_l}{2}t)}\sqrt{4\pi\nu}e^{\frac{C}{2\nu}u_l^{2/3}t^{2/3}}e^{ct^{1/3}},$$

where the last expression clearly goes to zero uniformly in ξ satisfying (3.37) (see (3.39)). Since for these ξ, D_t can be uniformly bounded from below by $c\sqrt{t}$, where $c > 0$ is an absolute constant, the proof is finished. $\square$

3.5 Front Probing and Central Limit Theorem

Now let us prove the main theorem of this chapter. The strategy is to study the solution of the regularized equation (3.10) along rays $\xi = \alpha t$ in the Hopf–Cole representation, and then take the limit $v \to 0$ to infer the same behavior for the Burgers solutions U. It follows from the two preceding propositions that we just need to study the limiting distribution of $\frac{B_t}{B_t/u_l+D_t}$.

Assume $\alpha > c$, so we are ahead of the actual front speed. In the ξ coordinate, this means that in the representation (3.21) of B_t, the factor $e^{(-u_l/2v)(\xi-u_l/2t)}$ goes exponentially fast to zero, uniformly in v. We now use the bound (3.28) (true with probability one for some constant C), and proceeding exactly as in the proof of Proposition 3.4, we have, with probability one,

$$\int_{\mathbb{R}} e^{-\frac{\eta^2}{4v} - \frac{\hat{x}(\xi-\sqrt{t}\eta-u_l t)}{2v}} \, d\eta \leq C e^{Ct^{2/3}}. \tag{3.40}$$

This clearly implies that $B_t \to 0$ almost surely as $t \to \infty$. On the other hand, $D_t \to +\infty$ (at the order of $\sqrt{t}$), so

$$\frac{B_t}{B_t/u_l + D_t} \to 0 \tag{3.41}$$

almost surely, and therefore the analogue of part 1 of the theorem is proven for the solution u_v, $v > 0$, solving the regularized equation (3.10). Note that all the above convergence statements, including (3.41), hold uniformly in v. It follows that

$$\sup_{v \leq v_0} |u_v(\xi(\alpha t),t)| \xrightarrow{d} 0. \tag{3.42}$$

Thanks to assumption A5, classical results [185, Theorems 13 and 14] and [207] yield that for any given t, except for a set of x consisting of countably many discontinuities of the first kind (shocks),

$$\lim_{v \to 0} u_v(x,t) = u_0(x,t). \tag{3.43}$$

Moreover, $u_0(x,t)$ is the unique physical weak solution of the (inviscid) Burgers equation, the so-called entropy solution [141]. It follows from (3.42) and (3.43) that

$$u_0(\xi(\alpha t),t) \xrightarrow{d} 0.$$

To prove the same convergence for U, note that by (3.11),

$$U(\alpha t,t) = \frac{1}{\sqrt{a(\alpha t)}} u_0(\xi(\alpha t),t). \tag{3.44}$$

Since the random variables $1/\sqrt{a(\alpha t)}$ have finite second moment uniformly in t (by A1 and A3), the product in (3.44) goes to zero in distribution. In fact, for any $\varepsilon > 0$ and $K \gg 1$,

$$P\left(\left|\frac{1}{\sqrt{a(\alpha t)}}u_0(\xi(\alpha t),t)\right| > \varepsilon\right) \leq P(|u_0(\xi(\alpha t),t)| > \varepsilon/K) + P\left(\frac{1}{\sqrt{a(\alpha t)}} > K\right),$$

where the first term goes to zero at large t, and the second term can be made as small as possible for large K by Chebyshev's inequality. This ends the proof of part one of the theorem, which says that we observe the value zero if we are ahead of the front.

Similarly, if $\alpha < c$, so we are behind the front, then the B_t term grows exponentially fast with probability one (due to the exponential prefactor in (3.21)), while D_t grows at most like $\sqrt{t}$ (if at all). Hence

$$\frac{B_t}{B_t/u_l + D_t} \to u_l, \tag{3.45}$$

and part two is proven for a positive v. Just as in the proof of part one, it suffices to note now that the convergence is uniform in v, and part two of the theorem follows. In contrast to part one, u_v does not converge to zero, and therefore we do not obtain convergence of $U(\alpha t,t)$. In view of (3.11), we see that $U(\alpha t,t)$ fluctuates as $t \to \infty$. So we observe a noisy state in U at the back of the front due to the effect of the random flux.

Now let us probe the front more precisely at

$$x = ct + z\sqrt{t}. \tag{3.46}$$

We want to find the distribution of

$$\frac{B_t}{B_t/u_l + D_t}$$

in the limit $t \to \infty$. The D_t integral behaves as $\sqrt{t}$ times a constant of order $O(\sqrt{v})$. Roughly speaking, B_t is either exponentially large or exponentially small. Accordingly, the above ratio is close to u_l or 0. This will be seen from the calculation below. Let $y \in (0, u_l)$ (note that $0 < B_t/\left(u_l^{-1}B_t + D_t\right) < u_l$). We have

$$P\left[\frac{B_t}{u_l^{-1}B_t + D_t} \leq y\right] = P\left[\frac{B_t}{D_t} \leq \frac{u_l y}{u_l - y}\right] = P\left[\frac{\log B_t}{\sqrt{t}} - \frac{\log D_t}{\sqrt{t}} \leq \frac{\log \frac{u_l y}{u_l - y}}{\sqrt{t}}\right]$$

$$= P\left[v\frac{\log B_t}{\sqrt{t}} - v\frac{\log D_t}{\sqrt{t}} \leq v\frac{\log \frac{u_l y}{u_l - y}}{\sqrt{t}}\right]. \tag{3.47}$$

Now, $v\frac{\log u_l y/(u_l - y)}{\sqrt{t}} \to 0$, and since D_t is of order $\sqrt{vt}$, we have $v\frac{\log D_t}{\sqrt{t}} \xrightarrow{d} 0$ as well. Both convergence statements hold uniformly in v in the sense explained in (3.26). The limit of the probability in (3.47) is therefore equal to

$$\lim_{t\to\infty} P\left[\frac{v\log B_t}{\sqrt{t}} \leq 0\right].$$

Now write B_t in product form:

$$B_t = p(t)\tilde{B}_t, \tag{3.48}$$

where $p(t) = e^{-\frac{u_l}{2\nu}(\xi - \frac{u_l}{2}t) - \frac{\hat{x}(\xi - u_l t)}{2\nu}}$ and

$$\tilde{B}_t = u_l \sqrt{t} \int_{\frac{\xi - u_l t}{\sqrt{t}}}^{\infty} e^{-\frac{\eta^2}{4\nu} + \frac{1}{2\nu}[\hat{x}(\xi - u_l t) - \hat{x}(\xi - u_l t - \sqrt{t}\eta)]} \, d\eta. \tag{3.49}$$

Changing the variable of integration, as in the proof of Proposition 3.4, to $y = t^{-1/6}\eta$, we obtain

$$\tilde{B}_t = t^{1/6} \int_{\frac{\xi - u_l t}{t^{2/3}}}^{\infty} e^{\frac{t^{1/3}}{2\nu}\left[\frac{\hat{x}(\xi - u_l t) - \hat{x}(\xi - u_l t - t^{2/3}y)}{t^{1/3}} - \frac{y^2}{2}\right]} \, dy. \tag{3.50}$$

Notice that

$$\frac{\hat{x}(\xi - u_l t) - \hat{x}(\xi - u_l t - t^{2/3}y)}{\mu^{-3/2}\sigma t^{1/3}}$$

converges in distribution to the Wiener process in the variable y, on any finite interval of y. Just as before, strengthening this to uniform convergence with a change of probability space (à la Skorohod) and applying the Laplace lemma, we find that as $t \to \infty$, the distribution of

$$t^{-1/3} \log \tilde{B}_t$$

converges to that of a constant times

$$\sup_y \left(\frac{y^2}{2} - W_y\right).$$

It follows that

$$\nu t^{-1/2} \log \tilde{B}_t \xrightarrow{d} 0$$

uniformly in ν, and we just need to study the behavior of $\nu t^{-\frac{1}{2}} \log p(t)$.

We have

$$\nu t^{-\frac{1}{2}} \log p(t) = -\frac{1}{2} \left[\frac{u_l(\xi - \frac{u_l}{2}t)}{\sqrt{t}} + \frac{\hat{x}(\xi - u_l t)}{\sqrt{t}}\right], \tag{3.51}$$

where $\xi = \xi(ct + z\sqrt{t})$. Since $c = 1/2\mu^2$ and $u_l = 1/\mu$, substituting (3.46), we get from the central limit theorem for ξ in assumption A4 that

$$\frac{u_l\left(\xi(ct + z\sqrt{t}) - \frac{u_l}{2}t\right)}{\sqrt{t}} \xrightarrow{d} z + \frac{\sigma}{\mu^2\sqrt{2}} W_1,$$

where W_1 is a Gaussian random variable with mean zero and unit variance. Also, using the central limit theorem for $\hat{x}$ ((3.24) with $b = 1/2\mu$), we obtain

$$\frac{\hat{x}(\xi - u_l t)}{\sqrt{t}} \xrightarrow{\text{d}} \frac{\sigma}{\mu^2 \sqrt{2}} W_1.$$

A further study of the joint distribution of the variables

$$\frac{u_l(\xi(ct + z\sqrt{t}) - \frac{u_l}{2}t)}{\sqrt{t}}$$

and

$$\frac{\hat{x}(\xi - u_l t)}{\sqrt{t}}$$

in the limit $t \to \infty$ shows that the two-dimensional random variables

$$\left(\frac{u_l(\xi(ct + z\sqrt{t}) - \frac{u_l}{2}t)}{\sqrt{t}}, \frac{\hat{x}(\xi - u_l t)}{\sqrt{t}} \right)$$

converge in distribution to a two-dimensional Gaussian with independent coordinates of means z and 0 respectively. The sum

$$\frac{u_l \left(\xi(ct + z\sqrt{t}) - \frac{u_l}{2}t \right)}{\sqrt{t}} + \frac{\hat{x}(\xi - u_l t)}{\sqrt{t}}$$

converges in distribution to a Gaussian random variable with mean z and variance σ^2/μ^4. Hence

$$P\left[vt^{-1/2} \log p(t) \leq 0 \right] \to P\left[-\frac{1}{2}(z + \frac{\sigma}{\mu^2} W_1) \leq 0 \right] = P\left[W_1 \geq -\frac{\mu^2}{\sigma} z \right]$$

$$= \mathcal{N}\left(\frac{\mu^2}{\sigma} z \right),$$

where $\mathcal{N}(s) = 1/\sqrt{2\pi} \int_{-\infty}^{s} e^{-s'^2/2} ds'$ is the error function. The above arguments can be made uniform in $v > 0$, implying that the Burgers solution u_0 satisfies

$$\lim_{t \to \infty} P\left[u_0(ct + z\sqrt{t}, t) \leq y \right] = \mathcal{N}\left(\frac{\mu^2}{\sigma} z \right)$$

and

$$\lim_{t \to \infty} P\left[u_0(ct + z\sqrt{t}, t) > y \right] = 1 - \mathcal{N}\left(\frac{\mu^2}{\sigma} z \right)$$

for all z, which can be rephrased as part 3 of the theorem on U.

3.6 Exercises

1. Prove the first assertion of the Laplace lemma, Lemma 3.2, by taking the limit as $\lambda \to \infty$ of the integral $\int_{\mathbb{R}^1} \psi(u)\,d\mu_\lambda$, where $\psi(u) \in C^\infty(\mathbb{R}^1)$, $|\psi(u)| \le C(1+u^2)^m$, for some $m > 0$. Show that the part of integral over a small neighborhood of the maximal point u_* can be made arbitrarily close to $\psi(u_*)$, and the remaining part of the integral is negligible for large λ. Part two of the Laplace lemma follows by setting $\psi(u) = u$.

2. Prove part three of the Laplace lemma, Lemma 3.2, by passing to the limit

$$\lambda^{-1} \log \int \exp\{\lambda(\varphi_\lambda(u) - \varphi(u_*))\}du \to 0, \quad \lambda \to \infty,$$

with a similar integral decomposition as in Exercise 1. Note that over a small neighborhood of the maximal point u_*, $\varphi_\lambda(u) - \varphi(u_*)$ is small for λ large enough.

3. Define the hitting time

$$T(b) = \inf\{x \ge 0 : \xi(x) = b\},$$

where ξ satisfies assumption A4. Express the probability of deviation of $T(tb)$ from tb/μ in terms of ξ to show that a law of large numbers holds for $T(tb)$:

$$\frac{T(tb)}{t} \overset{p}{\to} \frac{b}{\mu}, \quad t \to \infty,$$

where convergence is in probability [72].

4. Write $T(tb) - tb/\mu$ in terms of $\zeta(x) = \xi(x) - \mu x$ and apply A4 to show that for any fixed b, as $t \to \infty$,

$$\frac{T(tb) - tb/\mu}{\mu^{-3/2}\sigma\sqrt{t}} \overset{d}{\to} W_b.$$

Chapter 4
Fronts and Stochastic Homogenization of Hamilton–Jacobi Equations

Hamilton–Jacobi (HJ) equations modeling fronts in spatially random media are first-order nonlinear partial differential equations (PDEs) of the form

$$u_t + H(x, \omega, \nabla_x u) = 0, \quad x \in \mathbb{R}^N, \, N \geq 1, \tag{4.1}$$

where H, the Hamiltonian, is random in x (position) and nonlinear in $\nabla_x u$ (momentum). In case of one space dimension ($N = 1$), a scalar conservation law with a random flux as in Chapter 3 follows by formally taking the x derivative of (4.1):

$$v_t + (H(x, \omega, v))_x = 0,$$

where $v = u_x$. The Hamiltonian becomes the nonlinear flux function. Naturally, front speeds in scalar laws are related to HJ solutions. However, simple front solutions of the form $p \cdot x - c(p)t$ no longer exist in (4.1) due to the complexity of random dependence.

A front can be analyzed in a hyperbolic scaling limit. In other words, space and time are scaled the same. If the space scale is large, such as $O(1/\varepsilon)$ for $\varepsilon \ll 1$, then the time scale should also be $O(1/\varepsilon)$. On these space–time scales, a front is an (asymptotic) invariant. If the medium is homogeneous and we scale $x = \tilde{x}/\varepsilon$, $t = \tilde{t}/\varepsilon$, then the exact front solution is

$$u(x,t) = p \cdot x - H(p)t = \frac{1}{\varepsilon}\left(p \cdot \tilde{x} - H(p)\tilde{t}\right).$$

It follows that the front solution is invariant if we scale $\tilde{u} = \varepsilon u$, or $\tilde{u}(\tilde{x}, \tilde{t}) = p \cdot \tilde{x} - H(p)\tilde{t}$. For inhomogeneous media, one considers the initial value problem $u(x,0) = p \cdot x$ of (4.1) with planar initial data and then recovers the front speed by $-\lim_{t \to \infty} u(0,t)/t$. Performing the same scaling change of variables for the HJ equation (4.1), we have

$$\tilde{u}_{\tilde{t}}^\varepsilon + H\left(\tilde{x}/\varepsilon, \omega, \nabla_{\tilde{x}} \tilde{u}^\varepsilon\right) = 0, \tag{4.2}$$

J. Xin, *An Introduction to Fronts in Random Media*, Surveys and Tutorials in the Applied Mathematical Sciences 5, DOI: 10.1007/978-0-387-87683-2_4,
© Springer Science + Business Media, LLC 2009

with initial data $\tilde{u}^\varepsilon(\tilde{x}, 0) = p \cdot \tilde{x}$. The large-$t$ limit problem on u is the same as the small-ε limit problem (4.2) of $\tilde{u}^\varepsilon$, which is known as homogenization. Homogenization of HJ in the periodic setting was first studied in the 1980s [148]. More recent work has concentrated on the random setting, which we shall discuss in more detail.

The upshot is that under certain conditions on the Hamiltonian and randomness, solutions of (4.2) converge as $\varepsilon \downarrow 0$ to a limiting function $\bar{u}$ satisfying an averaged (homogenized) HJ equation

$$\bar{u}_{\tilde{t}} + \bar{H}(\nabla_{\tilde{x}} \bar{u}) = 0, \qquad (4.3)$$

with the same initial data. In the limiting process, the fast oscillations have been removed, and $\bar{H}$ is a deterministic function even if H is random to begin with. Homogenization is the analogue of the law of large numbers for stochastic PDEs. The homogenized equation (4.3) has the exact front solution

$$\bar{u}(\tilde{x}, \tilde{t}) = p \cdot \tilde{x} - \bar{H}(p)\tilde{t}. \qquad (4.4)$$

We observe at $(x, t) = (0, 1/\varepsilon)$ or $(\tilde{x}, \tilde{t}) = (0, 1)$ to obtain

$$\varepsilon u(0, 1/\varepsilon) = \tilde{u}^\varepsilon(0, 1) \rightarrow \bar{u}(0, 1) = -\bar{H}(p). \qquad (4.5)$$

Letting $T = 1/\varepsilon \gg 1$, (4.5) implies

$$u(0, T) \sim -\bar{H}(p)T, \qquad (4.6)$$

or the front speed exists asymptotically as $\bar{H}(p)$ in the direction p. Fronts in random media are therefore closely connected to the stochastic homogenization of HJ.

In this chapter, we shall discuss sufficient conditions for stochastic homogenization and related central limit theorems, then show examples in which homogenization breaks down. The latter (anomalous) regime corresponds to the study of extrema of stochastic sequences and processes [143] in probability theory. We shall give necessary and sufficient conditions for homogenization of the random Hamiltonians $H(x, \omega, p) = |p|^2/2 + V(x, \omega)$ in classical mechanics, and point out the consequences of extreme behavior of random media on the anomalous front dynamics.

4.1 Convex Hamilton–Jacobi and Variational Formulas

Let us consider the homogenization problem (4.2) with the tildes removed:

$$u_t^\varepsilon + H(x/\varepsilon, \omega, \nabla_x u^\varepsilon) = 0, \qquad (4.7)$$

where the superscript ε on u denotes its ε dependence. The major assumptions on the random Hamiltonian are these:

(A1)   Convexity: $H = H(x, \omega, p)$ is convex in p for any $(x, \omega) \in \mathbb{R}^N \times \Omega$. In other words, for any $p_1, p_2 \in \mathbb{R}^N$ and any $\theta \in [0, 1]$, we have for any (x, ω),

$$H(x,\omega,\theta p_1 + (1-\theta)p_2) \leq \theta H(x,\omega,p_1) + (1-\theta)H(x,\omega,p_2).$$

(A2)   Stationarity and ergodicity: H is a stationary and ergodic random field in (x,ω) for any p with respect to the shift (translation) in x.

(A3)   Coercivity: There exist positive deterministic constants C_1 and C_2 such that for some exponent $\alpha > 1$,

$$C_2\left(|p|^\alpha - 1\right) \leq H(x,\omega,p) \leq C_1\left(|p|^\alpha + 1\right). \tag{4.8}$$

(A4)   Continuity in x: There exists a continuous function $m : [0,\infty) \to [0,\infty)$ with $m(0) = 0$ such that

$$|H(x,\omega,p) - H(y,\omega,p)| \leq m(|x-y|)(1+|p|). \tag{4.9}$$

When m is a linear function, Lipschitz continuity holds and (A4) will be called (A4)' below.

A few words are in order for the assumptions. Convexity and coercivity help to define the Lagrangian function L via the Legendre transform

$$L(x,\omega,p) = \sup_{q \in \mathbb{R}^N} (p \cdot q - H(x,\omega,q)) = \max_{q \in \mathbb{R}^N} (p \cdot q - H(x,\omega,q)), \tag{4.10}$$

which is also convex and coercive. With the Lagrangian L and the Lipschitz regularity (A4)', HJ solutions are represented in terms of a variational path integral formula (the Lax formula [141, 147]):

$$u^\varepsilon(x,t,\omega) = \inf_{y \in \mathbb{R}^N} (g(y) + S^\varepsilon(x,y,y,\omega))$$

$$\equiv \inf_{y \in \mathbb{R}^N} \left(g(y) + \inf_{\xi \in A} \int_0^t L(\xi(s)/\varepsilon, \dot\xi(s), \omega)\, ds\right), \tag{4.11}$$

where the set A consists of all Lipschitz continuous paths $\xi(s)$ joining $y = \xi(0)$ to $x = \xi(t)$, and the function S^ε is called the action. The function g is the initial data of u^ε and is assumed to be Lipschitz continuous on $\mathbb{R}^N$. If H is x-independent (homogeneous media), then the Lax formula reduces to the Hopf formula [80]; see (4.13).

The ergodicity assumption in (A2) provides the averaging mechanism for homogenization. A stronger and more quantitative assumption than ergodicity is "mixing" [36, 72], which essentially means fast decay of correlations or independence of events that occur in vastly separate spatial regions. In case $N = 1$, x may also be thought of as time; then mixing says that the ancient past is nearly independent of the future. The mixing assumption is indirectly made in (A4) for the analysis of Burgers fronts in Chapter 2. Stationarity is also assumed in Chapter 2, and it makes the ensemble-averaged statistical quantities equal to constants instead of being x-dependent.

With stationarity and ergodicity assumptions, H can be generated by the operator τ_x (shift by x), with $H(x,\omega,p) = \tilde{H}(\tau_x\omega,p)$ for some function $\tilde{H}(\cdot,p)$ convex in p. Moreover, any shift-invariant set of events occurs with probability zero or one. Due to (A2), the averaged Hamiltonian is expected to be deterministic and x-independent. Assumption (A2) has appeared already in stochastic homogenization of linear elliptic PDEs [190].

The proof of homogenization will first utilize the Lax formula (4.11) to average out randomness and obtain the homogenized Lagrangian $\bar{L}$ and then recover $\bar{H}$ from the Legendre transform. The convergence in the mean is given by the following theorem [200]:

Theorem 4.1. *Assume (A1)–(A4) and that the initial datum g is Lipschitz, and consider (x,t) on a compact set D of $\mathbb{R}^N \times [\delta,\infty)$. Then for each $\delta > 0$, the quantity u^ε as given by the Lax formula (4.11) satisfies*

$$\lim_{\varepsilon\downarrow 0} E\left[\sup_D |u^\varepsilon(x,t,\omega) - \bar{u}(x,t)|\right] = 0, \tag{4.12}$$

where $\bar{u}$ is given by the Hopf formula of the homogenized equation (4.3):

$$\bar{u}(x,t) = \inf_{y\in\mathbb{R}^N}[g(y) + t\bar{L}((x-y)/t)], \tag{4.13}$$

where $\bar{L}$, the homogenized Lagrangian, is both convex and coercive. If Lipschitz continuity (A4)′ holds, then u^ε by the Lax formula satisfies the HJ equation (4.7).

Analogous assumptions in [200] are weaker than what we have stated here in that the constants C_1–C_3 are allowed to be random constants satisfying certain moment conditions, and that coercivity is more general. By the Hopf formula, $\bar{u}$ satisfies the homogenized HJ equation (4.3) almost everywhere [80].

Under suitable initial conditions, u^ε from the Lax formula is the unique viscosity solution of the initial value problem of HJ [147], a class of weak solutions from taking the zero viscosity limit as was done for Burgers' equation in Chapter 2. Compactness and semigroup property of viscosity solutions yields another mode of convergence up to the initial time [227]:

Theorem 4.2. *Assume (A1)–(A4). Then the unique viscosity solutions of (4.2) from bounded uniformly continuous initial data converge almost surely to those of (4.3) over compact sets of $\mathbb{R}^N \times [0,\infty)$.*

For a systematic exposition of viscosity solutions, see [147, 80].

In the sequel, we shall view Lax and Hopf formulas as generalized HJ solution formulas, and study the passage to the limit.

4.2 Subadditive Ergodic Theorem and Homogenization

The main part of the homogenization proof via the Lax formula is the application of the subadditive ergodic theorem [72] to the convergence of the action functional S^ε. Let us prove a theorem:

Theorem 4.3. *For fixed* $x, y \in \mathbb{R}^N$, *and* $t > 0$, *we have*

$$\lim_{\varepsilon \downarrow 0} E\left[|S^\varepsilon(x,y,t) - t\bar{L}((x-y)/t)|\right] = 0, \tag{4.14}$$

where $\bar{L}$ *is a convex and coercive function on* $\mathbb{R}^N$.

Proof. In preparation, we first observe a scaling property of S^ε:

$$S^\varepsilon(x,y,t,\omega) = \varepsilon S^1(x/\varepsilon, y\varepsilon, t\varepsilon, \omega). \tag{4.15}$$

This follows by a change of variables in (4.11), and is left as an exercise.
   The next observation is that

$$S^\varepsilon(x,y,t,\omega) \overset{\text{d}}{=} S^\varepsilon(x-y,0,t,\omega),$$

by stationarity of the random media. So it suffices to prove a special case of Theorem 4.3 that almost surely and in L^1 (in the sense of mean (4.14)), the scaled action functional $S^\varepsilon(x,0,t,\omega)$ converges to $t\bar{L}(x/t)$. To this end, let us consider a doubly indexed random sequence of numbers:

$$S_{m,n} = S^1(nx, mx, (n-m)t, \omega). \tag{4.16}$$

By the scaling property (4.15) and taking $\varepsilon = 1/n$, we see that $S^\varepsilon(x,0,t,\omega) = S_{0,n}/n$. The doubly indexed sequence essentially contains information on the "costs" of a trip from mx to nx in time $(n-m)t$. For the limit $S_{0,n}/n$ to be well behaved, the sequence $S_{m,n}$ needs to satisfy a subadditive property and finite moment conditions, which are summarized in Kingman's subadditive ergodic theorem [72, 200]:

Theorem 4.4. *Suppose that* $S_{m,n}$ *are random variables satisfying the following conditions:*

1. $S_{0,0} = 0$, $S_{m,n} \le S_{m,k} + S_{k,n}$, *for* $m \le k \le n$;
2. $\{S_{m,m+k}, m \ge 0, k \ge 0\}$ *equals* $\{S_{m+1,m+k+1}, m \ge 0, k \ge 0\}$ *in distribution;*
3. $E[S_{0,1}^+] < +\infty$; $\alpha_n \equiv E[S_{0,n}] < \infty$, $E S_{0,n} \ge \gamma_0 n$, *for a finite constant* γ_0.

   Then the following hold:

1.
$$\alpha = \lim_{n \to \infty} \frac{\alpha_n}{n} = \inf_{n \ge 1} \frac{\alpha_n}{n} \in (-\infty, \infty);$$

2. $S_\infty = \lim_{n \to \infty} S_{0,n}/n$ *exists with probability one;*

3. $\lim_{n \to \infty} E[|S_{0,n}/n - S_\infty|] = 0$.

Let us now verify that the sequence $S_{m,n}$ satisfies the three conditions in Kingman's theorem: subadditivity, stationarity, and finite moments. By definition of S^ε, the action $S^\varepsilon(x,y,t,\omega)$ is the infimum of a collection of Lipschitz paths connecting y to x in time t; hence making a stop in the middle only may increase the cost:

$$S^\varepsilon(x,y,t+s,\omega) = \inf_z [S^\varepsilon(x,z,t,\omega) + S^\varepsilon(z,y,t,\omega)] \leq S^\varepsilon(x,z,t,\omega) + S^\varepsilon(z,y,t,\omega),$$

for any z, implying subadditivity:

$$
\begin{aligned}
S_{m,n}(\omega) &= S^1(nx, mx, (n-m)t, \omega) \\
&\leq S^1(kx, mx, (k-m)t, \omega) + S^1(nx, kx, (n-k)t, \omega) \\
&= S_{m,k}(\omega) + S_{k,n}(\omega),
\end{aligned}
\tag{4.17}
$$

for any $m \leq k \leq n$. In simple terms, subadditivity says that the direct flight from m to n is cheaper than a one-stop flight, a consequence of the Lax formula.

For stationarity, we again infer from the definition of action (ξ being Lipschitz path) that

$$
\begin{aligned}
S_{m,n}(\omega) &= \inf \left\{ \int_0^{(n-m)t} L(\xi(s), \dot\xi(s), \omega) \, ds \,\Big|\, \xi(0) = mx, \xi((n-m)t) = nx \right\} \\
&= \inf \left\{ \int_0^{(n-m)t} L(\xi(s) + mx, \dot\xi(s), \omega) \, ds \,\Big|\, \xi(0) = 0, \xi((n-m)t) = (n-m)x. \right\} \\
&= \inf \left\{ \int_0^{(n-m)t} L(\xi(s), \dot\xi(s), \tau_{mx}\omega) \, ds \,\Big|\, \xi(0) = 0, \xi((n-m)t) = (n-m)x \right\} \\
&= S_{0,n-m}(\tau_{mx}\omega).
\end{aligned}
$$

Then it follows from the invariance of probability under translation τ that

$$
\begin{aligned}
\{S_{m,m+k}(\omega) : m,k \geq 0\} &= \{S_{0,k}(\tau_{mx}\omega) : m,k \geq 0\} \\
&\overset{\text{d}}{=} \{S_{0,k}(\tau_{(m+l)x}\omega : m,k \geq 0\} \\
&= \{S_{m+l,m+k+l}(\omega) : m,k \geq 0\}.
\end{aligned}
\tag{4.18}
$$

Coercivity and convexity of the Hamiltonian implies that of the Lagrangian [80], so there exists a convex and superlinearly increasing function $\psi(p)$ such that

$$\tilde{C}_2(\psi(p) - 1) \leq L(x, p, \omega) \leq \tilde{C}_1(\psi(p) + 1),$$

for positive constants $\tilde{C}_i$, $i = 1, 2$. It follows that for any $m \leq n$,

$$\tilde{C}_2(t\psi((x-y)/t) - t) \leq S^\varepsilon(x,y,t,\omega) \leq \tilde{C}_1(t\psi((x-y)/t) - t),$$

or

$$\tilde{C}_2((n-m)t\psi(x/t) - (n-m)t) \le S^1(nx, mx, (n-m)t, \omega)$$
$$\le \tilde{C}_1((n-m)t\psi(x/t) + (n-m)t).$$

Hence $|S_{0,n}|/n \le \tilde{C}_3$ for a positive constant $\tilde{C}_3$, or $S_{0,n}/n$ is bounded away from infinity uniformly in n and ω for fixed (x,t), $t > 0$.

The subadditive ergodic theorem, Theorem 4.4, applies to give

$$\lim_{n\to\infty} E[S_{0,n}/n] = \lim_{n\to\infty} S_{0,n}/n = \lim_{n\to\infty} S^{1/n}(x,0,t,\omega) \equiv \bar{S}(x,t), \qquad (4.19)$$

almost surely and in the mean. The limiting value is τ-invariant, and so is deterministic by ergodicity of the random media.

Though the above limit (4.19) is established for $\varepsilon = 1/n$, it can be extended to any ε thanks to the Lipschitz continuity of the action S^ε uniformly in (ε, ω) over compact sets of (x,t) $(t \ge \delta > 0)$.

The Lipschitz continuity of the action S^ε follows from (A1) and (A3); see [200, Lemma 3.1] for a proof. Note that assumption (A3) implies [200, inequality (3.17)], which is sufficient.

Now write $\varepsilon^{-1} = n + r$, $r \in (0,1)$. Then

$$\varepsilon S^1\left(\frac{x}{\varepsilon}, 0, \frac{t}{\varepsilon}, \omega\right) = \varepsilon S^1(nx + rx, 0, nt + rt, \omega) = \varepsilon n S^{1/n}\left(x + \frac{rx}{n}, 0, t + \frac{rt}{n}, \omega\right),$$

and by Lipschitz continuity,

$$\left| S^{1/n}\left(x + \frac{rx}{n}, 0, t + \frac{rt}{n}, \omega\right) - S^{1/n}(x, 0, t, \omega) \right| \le C_4 \frac{|x| + t}{n},$$

implying that

$$\lim_{\varepsilon \downarrow 0} \varepsilon S^1\left(\frac{x}{\varepsilon}, 0, \frac{t}{\varepsilon}, \omega\right) = \bar{S}(x,t),$$

almost surely and in the mean.

The scaling property of the action S^ε leads to the self-similar form of the limiting function $\bar{S}$:

$$\begin{aligned}
\lim_{\varepsilon\to 0} S^\varepsilon(x,0,t,\omega) &= \lim_{\varepsilon\to 0} \varepsilon S^1\left(\frac{x}{\varepsilon}, 0, \frac{t}{\varepsilon}, \omega\right) \\
&= \lim_{\varepsilon\to 0} (t\varepsilon) S^1\left(\frac{x}{t\varepsilon}, 0, \frac{t}{t\delta}, \omega\right) \\
&= t \lim_{\varepsilon\to 0} S^\varepsilon\left(\frac{x}{t}, 0, 1, \omega\right) \\
&= t\bar{S}\left(\frac{x}{t}, 1\right) \equiv t\bar{L}\left(\frac{x}{t}\right), \qquad (4.20)
\end{aligned}$$

where $\bar{L}$ is the homogenized Lagrangian function.

For general $y \ne 0$, we have

$$S^\varepsilon(x,y,t,\omega) = S^\varepsilon(x-y,0,t,\tau_{y/\varepsilon}\omega) \overset{\mathrm{d}}{=} S^\varepsilon(x-y,0,t,\omega),$$

by shift invariance of probability, and so

$$\lim_{\varepsilon \downarrow 0} E\left[|S^{\varepsilon}(x,y,t,\omega) - t\bar{L}((x-y)/t)|\right] = 0.$$

Let us show that $\bar{L}$ is both convex and coercive. By subadditivity, we have for any $a, b \in \mathbb{R}^N$, $\lambda \in [0, 1]$, that

$$S^{\varepsilon}(\lambda a + (1-\lambda)b, 0, 1, \omega) \leq S^{\varepsilon}(\lambda a, 0, \lambda, \omega) + S^{\varepsilon}(\lambda a + (1-\lambda)b, \lambda a, 1 - \lambda, \omega).$$

Then letting $\varepsilon \downarrow 0$ gives, in view of (4.20),

$$(1-\lambda)\bar{L}(b) = \lim_{\varepsilon \downarrow 0} S^{\varepsilon}(\lambda a + (1-\lambda)b, \lambda a, 1 - \lambda, \omega)$$

and the inequality

$$\bar{L}(\lambda a + (1-\lambda)b) \leq \lambda \bar{L}(a) + (1-\lambda)\bar{L}(b).$$

The convexity of $\bar{L}$ follows.

By coercivity of the Hamiltonian (A3) and the Legendre transform, there exist positive constants c_1 and c_0 depending only on α and the constants C_1, C_2 in (A3) such that

$$L\left(\frac{\xi}{\varepsilon}, \dot{\xi}, \omega\right) \geq c_1 \psi(\dot{\xi}) - c_0 t, \tag{4.21}$$

where $\psi(p) = |p|^{\beta}$, $\beta = \frac{\alpha}{\alpha-1}$ the conjugate exponent to α, and $\beta > 1$. It follows that

$$S^{\varepsilon}(x,y,t,\omega) = \inf_{\xi \in A} \int_0^t L\left(\frac{\xi}{\varepsilon}, \dot{\xi}, \omega\right) ds \geq c_1 t \psi((x-y)/t) - c_0 t. \tag{4.22}$$

Now letting $t = 1$, $y = 0$, and sending $\varepsilon \downarrow 0$, we arrive at the inequality

$$\bar{L}(x) \geq c_1 \psi(x) - c_0.$$

A similar upper bound $\bar{L}(x) \leq c_2 \psi(x) + c_3$ holds for positive constants c_2 and c_3, and so $\bar{L}$ satisfies the coercivity condition (A3).

This ends the proof of Theorem 4.3. $\square$

The convergence of u^{ε} in (4.11) requires boundedness of minimizers of $g(y) + S^{\varepsilon}(x,y,t,\omega)$ in addition to the convergence of S^{ε} in Theorem 4.3. Let us prove a lemma:

Lemma 4.5. *Fix $T > 0$. There exists a number $R = R(T)$ such that for $\forall t \in (0, T]$, and $\forall \varepsilon > 0$,*

$$\inf_{y \in \mathbb{R}^N} [g(y) + S^{\varepsilon}(x,y,t,\omega)] = \inf_{|y-x| \leq R(T)} [g(y) + S^{\varepsilon}(x,y,t,\omega)] \tag{4.23}$$

and

$$\inf_{y\in\mathbb{R}^N}[g(y)+t\bar{L}((x-y)/t)] = \inf_{|y-x|\leq R(T)}[g(y)+t\bar{L}((x-y)/t)]. \tag{4.24}$$

By coercivity of L, for any set $B \subset \mathbb{R}^N$, we have

$$\inf_{y\in B}[g(y)+tc_1\psi((x-y)/t)] - c_0 t \leq \inf_{y\in B}[g(y)+S^\varepsilon(x,y,t,\omega)]. \tag{4.25}$$

It follows by taking $y = x$ that

$$\inf_{y\in\mathbb{R}^N}[g(y)+S^\varepsilon(x,y,t,\omega)] \leq g(x)+S^\varepsilon(x,x,t,\omega) \leq g(x)+C_4 t \tag{4.26}$$

for a positive constant C_4. By the superlinear growth of ψ and at most linear growth of g, there exists R such that if $|x-y| \geq RT$, then

$$g(x)+C_4 t \leq g(y)+tc_1\psi((x-y)/t) - c_0 t. \tag{4.27}$$

Define $D = \{y : |x-y| \leq RT\}$. Then (4.25) implies

$$g(x)+C_4 t \leq \inf_{y\in D^c}[g(y)+tc_1\psi((x-y)/t)] - c_0 t. \tag{4.28}$$

Letting $B = D^c$ in (4.25), then combining (4.25), (4.26), and (4.28), we obtain, by applying the coercive bound on L,

$$\inf_{y\in\mathbb{R}^N}[g(y)+S^\varepsilon(x,y,t,\omega)] \leq \inf_{y\in D^c}[g(y)+S^\varepsilon(x,y,t,\omega)],$$

which proves (4.23). Inequality (4.24) follows similarly from the coercivity bounds of $\bar{L}$. Finally, Theorem 4.1 follows from Theorem 4.3, Lemma 4.5, and Lipschitz continuity of S^ε uniformly in ε.

Example 4.6. Consider the random Hamiltonian $H(x,\omega,p) = |p|^{2n} + V(x,\omega)$, $p \in \mathbb{R}^N$, where n is a positive integer, V is a stationary and ergodic random field; $|V| \leq C$ for all (x,ω) for some finite deterministic constant C, and V has continuous sample paths. All assumptions $(A1) - (A4)$ are satisfied.

Example 4.7. The Hamiltonian corresponding to the random flux function of the Burgers equation in Chapter 3 is $H(x,\omega,p) = a(x,\omega)p^2/2$, where x and $p \in \mathbb{R}^1$. For coercivity (A3) to hold, $a(x,\omega)$ must be bounded uniformly away from zero and infinity, or there are two positive deterministic constants C_1 and C_2 such that $C_1 \leq a(x,\omega) \leq C_2$. This is more restrictive than the positivity and moment assumptions in Chapter 3, where a is allowed to be a positive unbounded process such as an exponential of a stationary ergodic Gaussian process (log-normal process).

4.3 Unbounded Hamiltonians: Breakdown of Homogenization

We see from Example 4.7 above that coercivity assumption (A3) is not optimal and may omit unbounded processes often used in statistical modeling such as Gaussian and log-normal processes. These processes are widely used because they are intrinsically Gaussian and have fewer parameters to estimate from data than non-Gaussian processes. In this section, we shall analyze examples in which coercivity is not satisfied and the Hamiltonian functions are allowed to be unbounded (not uniformly bounded in x).

This turns out to be a rich territory where much more work is needed. Let us illustrate this point for specific examples. For simplicity, we shall consider only the existence of front speed or the limit

$$\bar{H}(p) \equiv -\lim_{\varepsilon \downarrow 0} u^{\varepsilon}(0, 1), \tag{4.29}$$

where u^{ε} is given by the Lax formula for front (affine) data $p \cdot x$.

The first example is the Hamiltonian in classical mechanics:

$$H(x, \omega, p) = |p|^2/2 + V(x, \omega), \tag{4.30}$$

where V is a stationary and ergodic random field with continuous sample paths and integrability

$$\mathbf{E}[|V(x)|] < \infty. \tag{4.31}$$

A precise result is given in the following theorem [74]:

Theorem 4.8. *For the random Hamiltonian* (4.30), *the one-sided almost sure upper bound* $V(x) \leq V_0$ *for a deterministic constant* V_0 *is necessary and sufficient for the existence of the HJ front speed (homogenized Hamiltonian). The limit* (4.29) *exists with probability one.*

In other words, the lower bound of the potential V is not necessary.

Let us verify the conditions of the subadditive ergodic theorem, Theorem 4.4. The Hamiltonian $H(x, p) = p^2/2 + V(x)$, a sum of kinetic and potential energy, describes a classical particle moving in the field of the potential V. The corresponding Lagrangian is $L(x, q) = |q|^2/2 - V(x)$. Fix $x \in \mathbb{R}^N$ and $t > 0$. The action integral for a particle moving from the origin to the point nx in time nt along the path $s \mapsto \xi(s)$, $0 \leq s \leq nt$, equals

$$\Phi_n = \int_0^{nt} L\left(\xi(s)\dot{\xi}(s)\right) ds.$$

We study the minimum $S_n(t, x)$ of Φ_n over all paths ξ such that $\xi(0) = 0$ and $\xi(nt) = nx$. Consider $S_{m,n}(t, x)$, where m and n are nonnegative integers such that $m < n$, defined as

$$S_{m,n}(t, x) = \min_{\substack{\xi(mt)=mx \\ \xi(nt)=nx}} \int_{mt}^{nt} L\left(\xi(t), \dot{\xi}(t)\right) dt, \tag{4.32}$$

with the minimum taken over all paths ξ connecting mx to nx in the time $(n-m)t$.

For $l < m < n$, we have clearly from (4.32) that

$$S_{0,m}(t,x) + S_{m,n}(t,x) \geq S_{0,n}(t,x).$$

Next, for a linear path $\xi(s) = sx/t$, we obtain

$$S_{0,n}(t,x) \leq \Phi_n = \frac{n|x|^2 t}{2} - \int_0^{nt} V\left(\frac{sx}{t}\right) ds,$$

which, by the integrability (4.31) of $V(x)$, implies that

$$\mathbf{E}[|S_{0,n}(t,x)|] < +\infty.$$

On the other hand, since the kinetic part of the action integral is positive, the upper bound on V implies that

$$E[S_{0,n}(t,x)] \geq -ntV_0.$$

The family $S_{m,n}$ thus satisfies assumptions (1)–(3) of the subadditive ergodic theorem, Theorem 4.4, implying the existence of the finite limit of $S_{0,n}/n$. This limit is invariant under translations of the realization of V by vectors proportional to x; hence its value is constant with probability one. Applying the scaling property of the action functional as before, one sees that the finite limit of $S_{0,n}/n$ equals $\bar{S}(x,t) = t\bar{L}((x-y)/t)$, where $\bar{L}$ is convex and is coercive from below:

$$\bar{L}(p) \geq \frac{|p|^2}{2} - V_0.$$

So $\bar{L}(p)$ grows superlinearly at large p. Moreover, the limit of the action holds for S^ε for any sequence of $\varepsilon \downarrow 0$ due to the Lipschitz continuity of S^ε, which follows from [200, Lemma 3.1]. Interestingly, the conditions of that lemma continue to hold when V is bounded from above.

To show that the asymptotic front speed exists, we modify Lemma 4.5 so that the boundedness of minimizers of $g(y) + S^\varepsilon(0, y, 1, \omega)$ holds almost surely. Because the right-hand side of inequality (4.26) is now $S^\varepsilon(0,0,\omega) = -V(0,\omega)$, the constant R is now a random constant to ensure that minimization occurs in $|y| \leq R(\omega)$. Then we pass $S^\varepsilon(0, y, 1, \omega)$ to its almost sure limit $\bar{L}(-y)$ and obtain the HJ asymptotic front speed

$$\bar{H}(p) = \inf_{y \in \mathbb{R}^N} [p \cdot y + \bar{L}(-y)]. \tag{4.33}$$

Now let us consider again a particle moving in a potential V that is a realization of a stationary random field, but this time the support of the distribution of $V(x)$ is unbounded from above. We also assume that the covariance of V decays sufficiently fast. We will show that unboundedness of V leads to a different behavior of the system, where the homogenization limit (4.29) fails because of the divergence of the action functional S^ε as $\varepsilon \to 0$.

It is instructive to look at the one-dimensional ($N = 1$) case, in which elementary calculations can be done. For simplicity, we take V to be the Ornstein–Uhlenbeck process [72], i.e., the Gaussian process with mean zero and the covariance function

$$\mathbf{E}[V(x)V(y)] = \frac{1}{2}\exp(-|x-y|).$$

Let us fix $x, t > 0$. For any n, the trajectory of the particle, starting from the point 0 at time $t = 0$ and arriving at the point nx at time nt, has classical action equal to

$$S_n = \min \int_0^{nt}\left[\frac{1}{2}\left(\frac{du}{d\tau}\right)^2 - V(u(\tau))\right] d\tau,$$

where the minimum is taken over all C^2 functions $u(\tau)$ such that $u(0) = 0$ and $u(nt) = nx$. Our goal is to study the asymptotic behavior of S_n/n as $n \to \infty$. Clearly, the function u that minimizes the action integral is monotone increasing and satisfies $\frac{du}{d\tau} \neq 0$ for all τ.

We can thus rewrite the action in terms of the minimum over the inverse functions $\tau(u)$:

$$S_n = \min \int_0^{nx}\left[\frac{1}{2}\left(\frac{d\tau}{du}\right)^{-1} - V(u)\frac{d\tau}{du}\right] du,$$

where the minimum is over all C^2 functions $\tau(u)$ such that $\tau(0) = 0$ and $\tau(nx) = nt$. It follows from the principle of energy conservation that the solution of the above variational problem satisfies

$$\frac{d\tau}{du} = \frac{1}{\sqrt{2[E_n - V(u)]}},$$

where E_n is the total energy of the particle and therefore

$$S_n = \int_0^{nx}\left[\sqrt{\frac{E_n - V(u)}{2}} - \frac{V(u)}{\sqrt{2(E_n - V(u))}}\right] du$$

and

$$\int_0^{nx}\frac{1}{\sqrt{2(E_n - V(u))}} du = nt.$$

In particular, $E_n > V(u)$ for all $u \in [0, nx]$ for the particle to be able to reach the point nx. We rewrite the formula for S_n as follows:

$$S_n = \int_0^{nx}\left[\sqrt{\frac{E_n - V(u)}{2}} + \frac{E_n - V(u)}{\sqrt{2(E_n - V(u))}} - \frac{E_n}{\sqrt{2(E_n - V(u))}}\right] du.$$

The first two terms in the integrand are identical, and their integral can be estimated by Jensen's inequality (using concavity of the root function):

$$\frac{1}{nx} \int_0^{nx} \sqrt{2(E_n - V(u))} \, du \leq \sqrt{\frac{1}{nx} \int_0^{nx} 2(E_n - V(u)) \, du}$$

$$= \sqrt{2E_n - 2\frac{1}{nx} \int_0^{nx} V(u) \, du}.$$

By the ergodicity of the Ornstein–Uhlenbeck process, we have

$$\frac{1}{nx} \int_0^{nx} V(u) \, du \to 0,$$

with probability 1 as $n \to \infty$, and therefore the integral of the first two terms in the expression for S_n is bounded by $3\sqrt{E_n} nx$ for large n. The integral of the third term is simply $-E_n nt$, so for large n we obtain

$$-tE_n \leq \frac{1}{n} S_n \leq 3x\sqrt{E_n} - tE_n.$$

As mentioned above, $E_n \geq \sup_{u \in [0,nx]} V(u)$. Since the realizations of the Ornstein–Uhlenbeck process $V(u)$, $u \geq 0$, are unbounded from above with probability one, it follows that $S_n / -ntE_n \to 1$ with probability 1. In particular, S_n/n is unbounded, which implies that the Hamilton–Jacobi equation with potential V does not homogenize in the usual sense [227, 200].

Moreover, since for $E_n > \sup_{u \in [0,nx]} V(u) + x^2/2t^2$ we have

$$\int_0^{nx} \frac{1}{\sqrt{2(E_n - V(u))}} \, du < nt,$$

it follows from the condition satisfied by E_n that $E_n \leq \sup_{u \in [0,nx]} V(u) + x^2/2t^2$. Consequently, S_n/n is asymptotically equivalent to $-tV^*(nx)$, where

$$V^*(y) = \sup_{u \in [0,y]} V(u)$$

is the running maximum of the Ornstein–Uhlenbeck process. The asymptotic equivalence of the two sequences means that their ratio converges to a positive constant.

We will invoke the following classical result about the running maximum of the Ornstein–Uhlenbeck process. It follows from the one-dimensional version of [4, Theorem 6.9.5].

Theorem 4.9. *The running maximum of $V^*(y)$ of the Ornstein–Uhlenbeck process satisfies the limit theorem*

$$\text{Prob} \left[\frac{V^*(y) - b_y}{a_y} \leq x \right] \to \exp(-e^{-x})$$

as $y \to \infty$, where

$$a_y^{-1} \sim b_y \sim (2 \log y)^{1/2}.$$

with $\sim$ denoting asymptotic equivalence.

The theorem says that the renormalized random variables $V^*(y)$ converge in distribution to the double exponential distribution. It follows that

$$\frac{V^*(y)}{(2\log y)^{1/2}} \to 1 \quad \text{as } y \to \infty$$

in probability, implying that

$$\frac{S_n}{n} \to -\infty$$

in probability.

The standard homogenization limit fails as claimed. Instead, the modified limit holds in probability:

$$\frac{S_n}{n(2\log n)^{1/2}} \to -t.$$

The divergence of homogenization means in particular that for affine data, the growth rate of HJ solutions in time is faster than linear. We shall come back to this point for front solutions in the next section.

Generalization to other random potentials is straightforward. In the above analysis, we have used only the fact that V is a mean-zero Gaussian process to which [4, Theorem 6.9.5] applies. For example, the divergence results hold for stationary Gaussian processes with the covariance function $r(\tau)$ satisfying

$$r(\tau) = 1 - C|\tau|^\alpha + o(|\tau|^\alpha)$$

as $\tau \to 0$ for some $\alpha \in (0,2]$, and $r(\tau)$ is square-integrable in $\tau \in \mathbb{R}^1$. The Ornstein–Uhlenbeck process satisfies this condition with $\alpha = 1$. Processes that satisfy it with $\alpha = 2$ have differentiable realizations.

The behavior of a particle in a multidimensional random potential unbounded from above is similar to that in one dimension. As before, we study the action S_n of the particle that goes from the origin to the point nx in time nt, where x is a fixed vector in $\mathbb{R}^d$. Let x_n^* denote the point in the ball $B(0, n|x|)$ where V reaches its maximum, equal to V_n^*. We also denote the minimum of V in the same ball by V_{*n}. To obtain an upper bound on S_n, consider a path that first goes from the origin to x_n^* in time δnt, moving with a constant velocity (equal to $x_n^*/\delta nt$), next spends time $(1-2\delta)nt$ at x_n^*, and finally goes from x_n^* to nx in the remaining time δnt, moving with the constant velocity $nx - x_n^*/\delta nt$. For such paths, we get

$$S_n \le \frac{1}{2}\delta nt \left(\frac{|x_n^*|^2}{\delta nt}\right)^2 - \delta nt V_{*n} - (1-2\delta)nt V_n^* + \frac{1}{2}\delta nt \left(\frac{|nx - x_n^*|}{\delta nt}\right)^2 - \delta nt V_{*n}.$$

It follows that

$$\frac{S_n}{n} \le \frac{5}{2}\frac{|x|^2}{\delta t} - 2\delta t V_{*n} - (1-2\delta)t V_n^*.$$

Because the process has mean zero, we have $V_{*n} = -V_n^*$ in law. Applying the asymptotic law of V_n^* [4, Theorem 6.9.5], we have the upper bound

$$\frac{S_n}{n} \leq \frac{5|x|^2}{\delta t} - \frac{1}{4} t V_{n|x|}^*,$$

for δ small enough. By almost sure logarithmic divergence of $V_{n|x|}^*$ as $n \to \infty$ (true for a Gaussian process), we conclude that $S_n/n \to -\infty$ almost surely, and the standard homogenization limit again fails. It will be interesting to establish a precise divergence result as in the case of one spatial dimension. However, the source of divergence is the same: dominance of the running maxima of an unbounded random process.

We see through the above case studies that the large space–time effects of randomness can take two very different forms:

1. Homogenization: In this case, the behavior of the system on large scales is described by an effective Lagrangian (or Hamiltonian) that is *nonrandom*. The disorder gets averaged, and the extreme nature of the random medium is tamed. An elementary (and linear) analogue of this phenomenon in classical probability theory is the strong law of large numbers [72].
2. Domination by finite-volume maxima of the random potential: In this case, homogenization breaks down, and on an arbitrarily large scale, the behavior of the system is dictated by the maximum value of the disorder on that scale. The extreme nature of the random medium prevails. A relevant problem of classical probability theory is the study of extrema of stochastic sequences and processes [143]. For example, the maximum M_n of n independent unit normal random variables behaves asymptotically as $\sqrt{2 \log n}$ and in particular, diverges as $n \to \infty$. Divergence of maxima of random fields underlies the breakdown phenomenon of stochastic homogenization.

The two types of behavior of random systems are incompatible with each other. For the class of Hamiltonians studied here, the behavior depends on a simple mathematical criterion: boundedness of the potential from above. The one-sided boundedness of the spatial potential is the sharp version of the boundedness assumption in the coercivity condition (A3).

4.4 Normal and Accelerated Fronts in Random Flows

Let us examine more examples of unbounded random Hamiltonians and interpret the results from the perspective of front speeds in random media.

Example 4.10 (Normal Fronts in Gradient Flows). Consider Hamiltonians of the form $H(x, \omega, p) = p^2/2 + p \cdot b(x, \omega)$, $p, x \in \mathbb{R}^d$. Here $b(x, \omega) = \nabla U(x, \omega)$, where $U(x, \omega)$ is a scalar random vector field whose realizations are of class C^2, and let us assume that the scalar random field

$$V_1(x,\omega) = -\frac{1}{2}|b|^2(x,\omega)$$

satisfies all the conditions on classical potentials in the last section (clearly, V_1 is bounded above). Examples of such fields b include Gaussian random fields with appropriate covariance [4]. The corresponding HJ equation models a front moving in a random gradient (compressible) flow field.

The Lagrangian function of the above H is

$$L(x,\omega,q) = \frac{1}{2}|q - b(x,\omega)|^2 = \frac{|q|^2}{2} - q \cdot b(x,\omega) + \frac{|b(x,\omega)|^2}{2}. \tag{4.34}$$

As in the previous section, consider a path $\xi(s), 0 \le s \le nt$ such that $\xi(0) = 0$ and $\xi(nt) = nx$. For such a path, the contribution from the second term to the Lagrangian integral is

$$\int_0^{nt} b(\xi(s)) \cdot \dot{\xi}(s)\,ds = U(nx).$$

This shows that this term is a null Lagrangian, i.e., that the Lagrangian (4.34) leads to the same Euler–Lagrange equations of motion as the Lagrangian of the potential system

$$L_1(x,q) = \frac{|q|^2}{2} - V_1(x,\omega),$$

where $V_1(x) = -\frac{1}{2}|b|^2(x,\omega)$. Thus Theorem 4.8 applies, and implies that *homogenization holds for such Hamiltonians of advection type, even though the flow field is unbounded.*

Example 4.11 (Front Divergence in Shear Flows). Let $b(x,\omega) = (V_2(x',\omega),0)$, $x' = (x_2,\dots,x_d)$, $0 \in \mathbb{R}^{N-1}$, $N \ge 2$, namely a shear flow in the direction x_1. The HJ equation associated with the Hamiltonian $H(x,\omega,p) = |p|^2/2 + b(x,\omega) \cdot p$ is

$$u_t + b(x,\omega) \cdot \nabla_x u + |\nabla_x u|^2/2 = 0. \tag{4.35}$$

Consider a solution that represents a front moving in the x_1 direction in the form $u(x,t) = x_1 - \frac{1}{2}t + w(x',t)$. Then w satisfies the HJ equation

$$w_t + |\nabla_{x'} w|^2/2 + V_2(x',\omega) = 0, \tag{4.36}$$

which is in the classical potential form. If V_2 obeys the assumptions in the previous section and is unbounded from above, then $-w(x',t)/t$ diverges for large time t, and the front speed is not asymptotically constant. Instead, there exists front speed acceleration due to the dominance of running maxima of the process V_2. Shear flow is a special incompressible (divergence-free) flow. It is interesting to study other types of random incompressible flows that may lead to divergence or anomalous behavior of front speeds.

The equation of motion for the Hamiltonian $H_1 = p^2/2 + b(x)p$, with $x, p \in \mathbb{R}^1$, is

$$\xi''(s) = b(\xi(s))b'(\xi(s)).$$

As discussed earlier, the same equation also arises from another Hamiltonian: $H_2 = p^2/2 + V_1(x)$, where $V_1(x) = -b^2(x)/2$. In this sense, the advection model is equivalent to a potential model, and homogenization of the two models is closely related.

The Lagrangians corresponding to H_1 and H_2 (their Legendre transforms) are respectively

$$L_1(x,q) = \frac{1}{2}(q - b(x))^2$$

and

$$L_2(x,q) = \frac{1}{2}(q + b^2(x)).$$

That the Hamiltonians H_1 and H_2 lead to the same equations of motion implies that the difference of the corresponding action functionals,

$$\int_0^{nt} \left[L_2(\xi(s), \dot{\xi}(s)) - L_1\left(\xi(s), \dot{\xi}(s)\right) \right] ds,$$

does not depend on the path $\xi(s)$. Taking the linear function $\xi(s) = \frac{x}{t}s$, we obtain

$$\int_0^{nt} \left[L_2\left(\xi(s), \dot{\xi}(s)\right) - L_1\left(\xi(s), \dot{\xi}(s)\right) \right] ds = \int_0^{nx} b(u)\, du.$$

Divided by n, this expression converges with probability one to the expected value of $b(0)$ times x. Consequently, homogenization or breakdown of homogenization holds for H_1 and H_2 simultaneously. In a general classical-mechanics Hamiltonian (as well as in a shear flow Hamiltonian), the potential can be unbounded from above. In contrast, the transformed potentials from the gradient flow Hamiltonians are nonpositive.

In conclusion, the mechanism behind the absence of homogenization in the random potential system is that the potential is unbounded from above and that the behavior of the system is dominated by the large fluctuations in the potential. In contrast, the Hamiltonian of unbounded random advection of gradient type is equivalent to a classical-mechanics Hamiltonian with a potential that though unbounded from below, is bounded from above, and so satisfies a homogenization principle. It is an interesting project to unravel more delicate conditions of stochastic HJ homogenization for other Hamiltonians in unbounded random media. A related problem is to study Hamiltonians with unbounded temporal fluctuations.

We shall see in the next chapter that analogous effective behavior is present for front speeds in reaction–diffusion–advection equations with unbounded random advection. So the phenomena we discussed in this chapter may have broader implications for wave propagation in random media. For reaction–diffusion fronts in incompressible random advection, temporal randomness is found to regularize the dominance of extreme events and promote mixing, and the speed of propagation is asymptotically a deterministic constant [247, 179, 182]. It is conceivable that similar results (or homogenization) hold for HJ equations in unbounded time-dependent random media under suitable conditions.

4.5 Central Limit Theorems and Front Fluctuations

Homogenization gives the leading-order asymptotic behavior $u^\varepsilon(x,t) \to \bar{u}(x,t)$, where the limiting function $\bar{u}$ is deterministic, analogous to the law of large numbers. More accurate is a stochastic approximation of u^ε, the central limit theorem (CLT), which carries information on statistics of fluctuations. We discussed such results in Chapter 3 on random Burgers fronts. In this section, we broaden our perspective in the context of convex HJ equations based on [200, 201], thereby extending CLT to fronts of convex conservation laws and HJs.

Let us consider HJ front solutions to (4.1) that travel with effective speeds

$$u(x,t,p) = w(x,p) - t\bar{H}(p), \qquad (4.37)$$

where $p \in \mathbb{R}^N$, and $w(x,p) = p \cdot x + o(|x|)$. These solutions are associated with the exact front solutions to the homogenized HJ (4.2), because in the scaling limit for homogenization, $\varepsilon w(x/\varepsilon, p) \to p \cdot x$. In the examples of this section such solutions can be found, and

$$w^\varepsilon(x,p) \equiv \varepsilon w\left(\frac{x}{\varepsilon}, p\right) \stackrel{d}{=} p \cdot x + \sqrt{\varepsilon} B(x,p) + o(\sqrt{\varepsilon}), \qquad (4.38)$$

for a random field $B(x,p)$ continuous in (x,p). Substituting (4.37) in (4.1) shows that $w(x,p)$ satisfies the time-independent HJ equation

$$H(x, \nabla_x w(x,p)) = \bar{H}(p). \qquad (4.39)$$

If we write $w(x,p) = p \cdot x + \hat{w}(x,p)$, then $\hat{w}$ solves the equation

$$H(x, p + \nabla_x \hat{w}) = \bar{H}(p), \qquad (4.40)$$

which is the corrector equation of homogenization, or the so-called cell problem in periodic homogenization.

Example 4.12. Let $H(x,p) = \frac{1}{2}a(x,\omega)p^2$, $x \in \mathbb{R}^1$, and $a(x,\omega) > 0$ as in Chapter 3. Then (4.40) implies $\bar{H} \geq 0$, and

$$p + \hat{w}_x = \sqrt{2\bar{H}}(a(x,\omega))^{-1/2}, \qquad (4.41)$$

or

$$\hat{w} = \hat{w}(0) + \int_0^x \frac{(2\bar{H})^{1/2}}{a^{1/2}(x')} \, dx' - px. \qquad (4.42)$$

Let us select $\bar{H}$ such that $\hat{w}(x,\omega)/x \to 0$ almost surely as $x \to \infty$. It follows that

$$p = (2\bar{H})^{1/2} \left\langle a^{-1/2} \right\rangle,$$

or

$$\bar{H}(p) = \frac{p^2}{2} \left\langle a^{-1/2} \right\rangle^{-2}. \tag{4.43}$$

Letting $p = 1$, this gives the asymptotic speed formula of Burgers fronts as in Chapter 3. By (4.43), (4.42) reads

$$\hat{w}(x) = p \left\langle a^{-1/2} \right\rangle^{-1} \int_0^x a^{-1/2}(x') \, dx' - px, \tag{4.44}$$

where without loss of generality we have set the constant term $\hat{w}(0)$ to zero. Applying the invariance principle (A4 in Chapter 3) to (4.44), we find that $w(x,p)$ is expressed as

$$w(x,p) \overset{d}{=} px + p\sigma \left\langle a^{-1/2} \right\rangle^{-1} W_x + o(\sqrt{x}), \tag{4.45}$$

implying (4.38) in the scaled form with $B(x,p)$ being p times a Wiener process in x.

Suppose (4.38) holds. Let us formally verify the asymptotic expansion

$$u^\varepsilon(x,t) \overset{d}{=} \bar{u}(x,t) + \sqrt{\varepsilon} \inf_{y \in I(x,t)} [B(x,p(x,y,t)) - B(y,p(x,y,t))] + o(\sqrt{\varepsilon}), \tag{4.46}$$

where $I(x,t)$ is the set of minimizers y of the Hopf formula (4.13) and

$$p(x,y,t) = \nabla \bar{L}((x-y)/t). \tag{4.47}$$

Let us find an asymptotic expansion of the action function $S^\varepsilon(x,y,t)$. For front solutions $u^\varepsilon(x,t,p)$, we have

$$u^\varepsilon(x,t,p) = \inf_y [w^\varepsilon(y,p) + S^\varepsilon(x,y,t)], \tag{4.48}$$

where u^ε is the solution of $u_t^\varepsilon + H(x/\varepsilon, \nabla_x u^\varepsilon) = 0$, $u^\varepsilon(x,0) = w^\varepsilon(x,p)$. If the infimum is attained at some $y^\varepsilon(x,t,p)$, we obtain

$$S^\varepsilon(x,y^\varepsilon(x,t,p),t) = u^\varepsilon(x,t,p) - w^\varepsilon(y^\varepsilon(x,t,p),p). \tag{4.49}$$

Suppose that the function $y = y^\varepsilon(x,t,p)$ is invertible for each (x,t), and denote the inverse function by $p = p^\varepsilon(x,y,t)$, or

$$y^\varepsilon(x,t,p^\varepsilon(x,y,t)) = y. \tag{4.50}$$

Since u^ε converges to $\bar{u}(x,t,p) = x \cdot p - t\bar{H}(p)$, equal to

$$\bar{u}(x,t,p) = \inf_y [y \cdot p + t\bar{L}((x-y)/t)], \tag{4.51}$$

the minimizer y satisfies $p = p(x,y,t) = \nabla \bar{L}((x-y)/t)$. We anticipate a CLT expansion for $p^\varepsilon(x,y,t)$ (or y^ε) to come from that of u^ε:

$$p^{\varepsilon}(x,y,t) = p(x,y,t) + \sqrt{\varepsilon}R(x,y,t) + o(\sqrt{\varepsilon}), \tag{4.52}$$

for a random process $R(x,y,t)$. It follows from (4.37)–(4.38) that

$$u^{\varepsilon}(x,t,p) = x \cdot p - t\bar{H}(p) + \sqrt{\varepsilon}B(x,p) + o(\sqrt{\varepsilon}). \tag{4.53}$$

Now we write (4.49) in terms of the inverse function p^{ε} and expand with the help of (4.52) and (4.53) as

$$\begin{aligned}
S^{\varepsilon}(x,y,t) &= u^{\varepsilon}(x,t,p^{\varepsilon}(x,y,t)) - w^{\varepsilon}(y,p^{\varepsilon}(x,y,t)) \\
&= [(x-y) \cdot p(x,y,t) - t\bar{H}(p(x,y,t))] \\
&\quad + \sqrt{\varepsilon}[(x-y) - t\bar{H}(p(x,y,t))] \cdot R(x,y,t) \\
&\quad + \sqrt{\varepsilon}[B(x,p(x,y,t)) - B(y,p(x,y,t))] + o(\sqrt{\varepsilon}).
\end{aligned} \tag{4.54}$$

One checks by the Legendre transform that the first term on the right-hand side of (4.54) equals $t\bar{L}((x-y)/t)$, and the second term is zero. It follows that

$$S^{\varepsilon}(x,y,t) = t\bar{L}((x-y)/t) + \sqrt{\varepsilon}[B(x,p(x,y,t)) - B(y,p(x,y,t))] + o(\sqrt{\varepsilon}). \tag{4.55}$$

Substituting (4.55) in the Lax formula (4.11) for a more general solution $u^{\varepsilon}(x,t)$ leads to the form of CLT expansion (4.46)–(4.47), where y is determined by the infimum of the sum of the $O(1)$ terms. The cancellation of the second term in (4.54) suggests that CLT expansion of p^{ε} is not necessary: a low-order expression of $p^{\varepsilon} = p + o(1)$ suffices. Because the CLT expansions rely on knowledge of $w(x,p)$, which may not exist in general [149], we state below two concrete results adapted from [201].

Theorem 4.13. *Suppose $H(x,p,\omega) = a(x,\omega)K(p)$, $x \in \mathbb{R}^1$, $a(x,\omega)$ is bounded between two deterministic positive constants with probability one, and $K : \mathbb{R}^1 \to [0,\infty)$ is a coercive, convex, and continuously differentiable function with $K(0) = 0$. Moreover, $a(x,\omega)$ satisfies stationarity and ergodicity for homogenization. The function K has two branches of inverses, $A^+ : [0,\infty) \to [0,\infty)$, $A^- : [0,\infty) \to (-\infty,0]$, so that $K(A^{\pm}(\lambda)) = \lambda \geq 0$. Let $\varphi^{\pm}(\lambda) = E[A^{\pm}(\lambda/a(0,\omega))]$. Define*

$$\bar{H}(p) = (\varphi^+)^{-1}(p)\chi(p \geq 0) + (\varphi^-)^{-1}(p)\chi(p < 0) \tag{4.56}$$

and

$$w(x,p) = \chi(p \geq 0)\int_0^x A^+(\bar{H}(p)/a(y,\omega))\,dy + \chi(p < 0)\int_0^x A^-(\bar{H}(p)/a(y,\omega))\,dy. \tag{4.57}$$

Then

1. $w(x,p)$ and $w_x(x,p)$ are continuous in (x,p);
2.

$$\lim_{p \to \infty} H_p(x,w_x(x,p)) = \pm\infty,$$

uniformly in x;

3. $H(x, w_x(x, p)) = \bar{H}(p)$, *for all* (x, p).

Assume additionally that $\bar{H}$ *is strictly convex (no degenerate linear pieces), and that the process* $\frac{1}{\sqrt{\varepsilon}}(\varepsilon w(x/\varepsilon, p) - p \cdot x)$ *converges to a continuous process* $B(x, p)$. *Then any finite-dimensional distributions of the process (error of homogenization)*

$$\gamma^\varepsilon(x, t, \omega) = \varepsilon^{-1/2}(u^\varepsilon(x, t, \omega) - \bar{u}(x, t))$$

converge to those of the process

$$\gamma(x, t) = \inf_{y \in I(x,t)}[B(x, p(x, y, t)) - B(y, p(x, y, t))], \qquad (4.58)$$

where $p = p(x, y, t) = \bar{L}'((x - y)/t)$, *the prime denoting the derivative and* $\bar{L}$ *being the Legendre transform of* $\bar{H}$.

The additional convergence assumption in the above theorem holds in the context of the invariance principle under the sufficient mixing conditions of $a(x, \omega)$, as in Chapter 3 for Burgers equations. As a result, the asymptotics (4.38) is valid and $B(x, p)$ is a Gaussian process in x, implying that the random variable $\gamma(0, 1)$ is Gaussian. Hence the front speed fluctuation of convex HJ in positive multiplicative one-dimensional random media obeys Gaussian statistics.

Next we consider the Hamiltonians of the type in classical mechanics:

$$H(x, p, \omega) = K(p) + V(x, \omega), \quad x \in \mathbb{R}^1, \qquad (4.59)$$

where $V(x, \omega)$ is a stationary and ergodic process for homogenization, and $-\infty < \inf_x V(x, \omega) \leq 0$ with probability one; $K : \mathbb{R}^1 \to [0, \infty)$ is a continuously differentiable strictly convex coercive function with $K(0) = 0$. So $K(p)$ is monotone increasing (decreasing) for $p > 0$ ($p < 0$). There are two branches of inverse functions $A^\pm$. Set $\varphi^\pm(\lambda) = E[A^\pm(\lambda - V(0, \omega))]$, both strictly monotone for $\lambda \geq 0$. Define

$$\bar{H}(p) = \chi(p \leq \varphi^-(0))\left(\varphi^-\right)^{-1}(p) + \chi(p \geq \varphi^+(0))\left(\varphi^+\right)^{-1}(p), \qquad (4.60)$$

and $\bar{L}$ its Legendre transform (homogenized Lagrangian). One may check that if $z \neq 0$, then $\bar{L}$ is differentiable at z with value $\bar{L}'(z) \in [\varphi^-(0), \varphi^+(0)]$.

Theorem 4.14. *Assume that the processes*

$$B_\pm^\varepsilon(x, \lambda, \omega) \equiv \varepsilon^{-1/2}\left(\varepsilon \int_0^{x/\varepsilon} A^\pm(\lambda - V(y, \omega))\,dy - x\varphi^\pm(\lambda)\right) \qquad (4.61)$$

converge in distribution to a continuous process $B^\pm(x, \lambda)$ *for any* $(x, \lambda) \in \mathbb{R}^1 \times [0, \infty)$. *Define*

$$B(x, p) = \chi(p \geq \varphi^+(0))B^+(x, (\varphi^+)^{-1}(p)) + \chi(p \leq \varphi^-(0))B^-(x, (\varphi^-)^{-1}(p)). \qquad (4.62)$$

Assume also that with probability one,

$$\lim_{n \to \infty} \frac{1}{n^2} \int_{-n}^{n} \frac{dz}{K'(A^\pm(-V(z,\omega)))} = \pm\infty. \tag{4.63}$$

Then any finite-dimensional distributions of the process (error of homogenization)

$$\gamma^\varepsilon(x,t,\omega) = \varepsilon^{-1/2}(u^\varepsilon(x,t,\omega) - \bar{u}(x,t)),$$

converge to those of the process

$$\gamma(x,t) = \inf_{y \in I(x,t) - \{x\}} [B(x,p(x,y,t)) - B(y,p(x,y,t))], \tag{4.64}$$

where $p = p(x,y,t) = \bar{L}'((x-y)/t)$.

Because of the degeneracy of $\bar{H}(p)$, namely $\bar{H}(p) = 0$ for $p \in [\varphi^-(0), \varphi^+(0)]$, the function $w(x,p)$ is complicated. The condition (4.63) allows one to use only $w(x,p)$ for p outside of $[\varphi^-(0), \varphi^+(0)]$ and a solution formula of the form

$$w(x,p) = \int_0^x A^\pm(\bar{H}(p) - V(y,\omega))\, dy$$

for analysis. This formula is used in (4.61). In the quadratic Hamiltonian case, $K(p) = p^2/2$, condition (4.63) becomes

$$\lim_{n \to \infty} \frac{1}{n^2} \int_{-n}^{n} \frac{dz}{\sqrt{\sup V(z,\omega) - V(z,\omega)}} = +\infty,$$

which is satisfied if V is twice differentiable and so has the asymptotics $V(z,\omega) - V(z_0,\omega) = O(|z - z_0|^2)$ near a local maximal point z_0 that belongs to $(-n,n)$ for n large enough.

Extensions of the above results to several spatial dimensions are possible if the existence of $w(x,p)$ and the asymptotic property (4.38) are established. This is a challenging problem by itself. Some special cases are known, for example $H(x,p,\omega) = \bar{H}((b(x,\omega))^{-1}(p - c(x,\omega)))$, where $\bar{H}$ is a strictly convex continuously differentiable and coercive function, b is an invertible matrix function, $c(x)$ is a vector function, and both b and c are in gradient form; see [200, 201].

4.6 Exercises

1. Generalize Theorem 4.8 to the random Hamiltonian

$$H(x,p,\omega) = |p|^{2n} + V(x,\omega),$$

where n is a positive integer.

2. Suppose that the action function $S^\varepsilon(x,y,t)$, $(x,y,t) \in \mathbb{R}^{2N} \times (0,+\infty)$, is continuous in (x,y,t) for any ε and almost all ω, and that for a convex function $\bar{L}$, the process

$$Z^\varepsilon(x,y,t,\omega) = \varepsilon^{-1/2}(S^\varepsilon(x,y,t,\omega) - t\bar{L}((x-y)/t))$$

converges to a continuous process $Z(x,y,t,\omega)$. Use the Lax formula to show that finite-dimensional distributions of

$$\gamma^\varepsilon(x,t,\omega) = \varepsilon^{-1/2}(u^\varepsilon(x,t,\omega) - \bar{u}(x,t)),$$

converge to those of

$$\inf_{y \in I(x,t)} Z(x,y,t),$$

where I is the set of minimizers in the Hopf formula for $\bar{u}(x,t)$.

3. Let $H(x,p)$ be convex in p and continuously differentiable in $(p,x) \in \mathbb{R}^{2N}$, and L its Legendre transform in p. Suppose that w is a continuously differentiable solution of

$$H(x, \nabla_x w(x)) = \lambda,$$

for some constant λ, and that $x(t)$ is a solution of

$$\frac{dx}{dt} = H_p(x, \nabla_x w(x)),$$

where H_p is the gradient of H with respect to p. Show that:

a. $L(x, H_p(x,p)) = p \cdot H_p(x,p) - H(x,p)$.
b.
$$\int_0^t L(x(s), \dot{x}(s)) ds = w(x(t)) - w(x(0)) - \lambda t.$$

c. For any $t > 0$,
$$w(a) - \lambda t = \inf_y[w(y) + S(a,y,t)] \qquad (4.65)$$

and

$$S(x(t), x(0), t) = \int_0^t L(x(s), \dot{x}(s)) ds = w(x(t)) - w(x(0)) - \lambda t. \qquad (4.66)$$

4. Under the assumptions of Theorem 4.13:

a. Apply (4.65)–(4.66) to show that

$$S(x,y,t,\omega) = \sup_p[w(x,p,\omega) - w(y,p,\omega) - t\bar{H}(p)], \qquad (4.67)$$

where $x = x(t,y,p)$ is the unique solution of

$$\frac{dx}{dt} = H_p(x, w_x(x,p)), \qquad (4.68)$$

$x(0, y, p) = y.$

b. Use (4.67) and that $\lim_{p \to \pm\infty} H_p(x, w_x(x, p)) = \pm\infty$ to show that

$$\lim_{p \to \pm\infty} x(t, y, p) = \pm\infty.$$

c. Deduce from part b above and (4.68) that for each (x, y, t) there exists a p such that $x(t, y, p) = x$. Hence (4.67) holds for all (x, y, t).

Chapter 5
KPP Fronts in Random Media

In this chapter, we discuss KPP fronts in space and/or time random flows by a combination of PDE and probabilistic methods. We first consider fronts in spatial random shear flows in a channel domain with finite width L. Here randomness appears in the transverse direction of the front. The variational principle (2.52) under zero Neumann boundary condition applies to each random speed, and allows a combined PDE–probabilistic analysis and a computational study of the speed ensemble. We identify new random phenomena such as the resonance in speed dependence on correlation length of the flows, speed slowdown due to temporal decorrelation of random flows, and speed divergence due to extreme behavior of random media. For example, for channel domain width $L \gg 1$, front speed diverges due to unbounded running maxima of the shear flow, which shares the same source of divergence as stochastic homogenization of Hamilton–Jacobi equations in Chapter 4. We shall present front growth or decay laws in random media, in comparison with those of periodic media in Chapter 3. Finally, we outline a recent breakthrough in solving the turbulent front speed problem for the KPP model, and study related speed bounds.

5.1 KPP Fronts in Spatially Random Shear Flows

Let us consider KPP front speeds through random shear flows in channel domains $D \equiv \mathbb{R} \times \Omega$, where $\Omega \subset \mathbb{R}^{n-1}$, $n \geq 2$, is a bounded simply connected domain with a smooth boundary. We address the enhancement properties of the ensemble-averaged front speeds, and their dependence on the flow statistics. The materials are based on [175, 178]. Recall the advection–diffusion equation with KPP reaction

$$u_t = \Delta_{x,y} u + B \cdot \nabla_{x,y} u + f(u), \tag{5.1}$$

where $t \in \mathbb{R}^+$, $\Delta_{x,y}$ the n-dimensional Laplacian, $(x, y) \in D$. The vector field $B = (b(y, \omega), \mathbf{0})$, where $b(y, \omega)$ is a stationary continuous scalar random process in y, has its ensemble mean equal to zero. The zero Neumann boundary condition is imposed

J. Xin, *An Introduction to Fronts in Random Media*, Surveys and Tutorials in the Applied Mathematical Sciences 5, DOI: 10.1007/978-0-387-87683-2_5,

at $\partial\Omega$: $\frac{\partial u}{\partial v} = 0$, v the unit outward normal. The KPP speed c^* variational formula (2.52) in Chapter 2 simplifies to

$$c^* = c^*(\omega) = \inf_{\lambda > 0} \frac{\mu(\lambda, \omega)}{\lambda}, \qquad (5.2)$$

where $\mu(\lambda, \omega)$ is the principal eigenvalue with corresponding eigenfunction $\phi > 0$ of the problem

$$\bar{L}_\lambda \phi = \Delta_y \phi + [\lambda^2 + \lambda b(y, \omega) + f'(0)]\phi = \mu(\lambda, \omega)\phi, \quad y \in \Omega, \qquad (5.3)$$

$$\frac{\partial \phi}{\partial v} = 0, \quad y \in \partial\Omega. \qquad (5.4)$$

We shall use the superscript $*$ to denote speed in random media in this chapter, which is different from the speed c_* in a deterministic medium. The derivation of (5.2)–(5.4) from (2.52) is left as an exercise.

5.1.1 Asymptotics of Averaged Speeds

Let us first study the deterministic case in which b in equation (5.3) is a continuous function with $\int_\Omega b(y)\,dy = 0$. We shall see what quantities appear in estimating the front speed, then extend them properly to the random setting.

Proposition 5.1. *Let $\chi = \chi(y)$ solve $\Delta_y \chi = -b$, $y \in \Omega$, with zero Neumann boundary condition, where $b \in C(\overline{\Omega})$ has zero integral over Ω. Then for δ sufficiently small, the minimal speed has the expansion*

$$c_*(\delta) = c_0 + \frac{c_0 \delta^2}{2|\Omega|} \int_\Omega |\nabla \chi|^2 \, dy + O(\delta^3). \qquad (5.5)$$

We shall give a proof of Proposition 5.1 using variational formulas, and later generalize it to the random case. A helpful fact is that the infimum in (5.2) can be restricted to a bounded set independent of b and δ, as stated in the following lemma:

Lemma 5.2. *Let $b \in C(\overline{\Omega})$ have zero mean over Ω, and let $\lambda_0 = \sqrt{f'(0)}$. Then*

$$\inf_{\lambda > 0} \frac{\mu(\lambda)}{\lambda} = \inf_{0 < \lambda \leq \lambda_0} \frac{\mu(\lambda)}{\lambda}. \qquad (5.6)$$

Proof. For each $c > 0$, we let $\rho_c(\lambda) = \mu(\lambda) - \lambda c - \lambda^2$. So if $\phi > 0$ is the eigenfunction defined by (5.3), then $\rho_c(\lambda)$ is the principal eigenvalue defined by the equation

$$\Delta_y \phi + [\lambda b(y) - \lambda c + f'(0)]\phi = \rho_c(\lambda)\phi, \quad y \in \Omega.$$

One can readily verify that $\partial_\lambda \rho_c(\lambda)|_{\lambda=0} = -c < 0$. The variational formula (5.2) can be expressed as

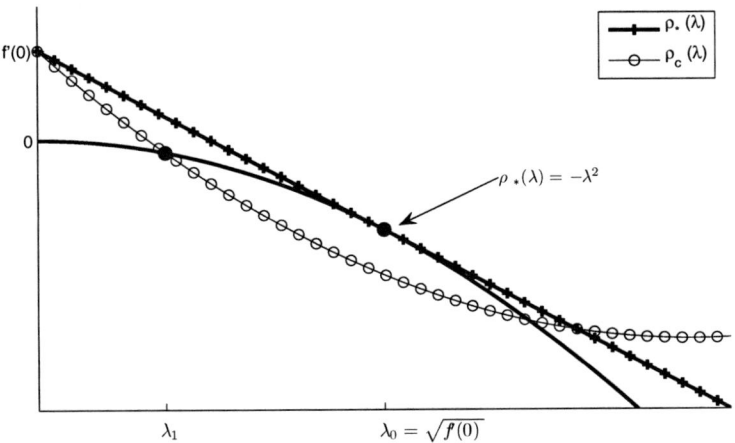

Figure 5.1 Intersecting curves ρ_c and $-\lambda^2$, from [178].

$$c^* = \inf\{c : \exists \lambda > 0, \lambda c = \mu(\lambda)\} = \inf\{c : \exists \lambda > 0, \rho_c(\lambda) = -\lambda^2\}. \qquad (5.7)$$

Consider the points where $\rho_c(\lambda) = -\lambda^2$. By [31, Proposition 2.1], the continuous curve $\lambda \mapsto \rho_c(\lambda)$ is convex in λ, for each $c > 0$. Also, $\rho_c(0) = f'(0) > 0$. Therefore, for a given $c > 0$, there can be at most two values of $\lambda > 0$ such that $\rho_c(\lambda) = -\lambda^2$. The line $\rho_*(\lambda) = -2\sqrt{f'(0)}\lambda + f'(0)$ satisfies $\rho_*(\lambda) \geq -\lambda^2$, with equality holding only at one point: $\lambda_0 = \sqrt{f'(0)}$. Since $\rho_*(0) = \rho_c(0)$ and $\rho_c(\lambda)$ is convex and ρ_* is a line, $\rho_c(\lambda) = -\lambda^2$ for some $\lambda > 0$ only if $\rho_c(\lambda_1) = -\lambda_1^2$ for some $\lambda_1 \in (0, \lambda_0]$. This point is illustrated in Figure 5.1.1. The solid curve represents the parabola $-\lambda^2$. If $\rho_c(\lambda)$ intersects $-\lambda^2$, then one of the intersection points must be to the left of $\lambda_0 = \sqrt{f'(0)}$. Therefore, from (5.7),

$$\{c : \exists \lambda > 0, \rho_c(\lambda) = -\lambda^2\} = \{c : \exists \lambda \in (0, \lambda_0], \rho_c(\lambda) = -\lambda^2\}.$$

So we conclude that

$$c^*(\delta) = \inf_{0 < \lambda} \frac{\mu(\lambda)}{\lambda} = \inf_{0 < \lambda \leq \lambda_0} \frac{\mu(\lambda)}{\lambda}. \qquad (5.8)$$

This completes the proof $\square$

Now we turn to the proof of Proposition 5.1.

Proof (of Proposition 5.1). To estimate $c^*(\delta)$, we bound the principal eigenvalue $\mu(\lambda)$ using two different representations of μ. First, since $\bar{L}$ is a self-adjoint operator, we have

$$\mu = \sup \frac{(\bar{L}_\lambda \psi, \psi)}{\|\psi\|_2^2}, \tag{5.9}$$

where the supremum is taken over all $\psi \in H^2(\Omega)$ such that $\frac{\partial \psi}{\partial \nu} = 0$ on $\partial \Omega$. The other representation is

$$\mu = \inf_\psi \sup_{y \in \Omega} \frac{\bar{L}_\lambda \psi}{\psi} = \inf_\psi \sup_{y \in \Omega} \left(\frac{\Delta \psi}{\psi} + \lambda \delta b + \lambda^2 + f'(0) \right), \tag{5.10}$$

where the infimum can be taken over all $\psi \in C^1(\Omega)$ such that $\Delta \psi \in C(\Omega)$, $\psi > 0$, and $\frac{\partial \psi}{\partial \nu} = 0$ on $\partial \Omega$. This representation follows from the fact that the eigenfunction $\phi > 0$ lies in the kernel of the self-adjoint operator $(\bar{L}_\lambda - \mu(\lambda)I) = (\bar{L}_\lambda - \mu(\lambda)I)^*$. So, if we have the strict inequality

$$\bar{L}_\lambda \psi - \mu(\lambda)\psi = m < 0, \tag{5.11}$$

then the Fredholm alternative implies that $(\phi, m)_{L^2} = 0$, a contradiction, since $\phi > 0$, $m < 0$. Hence

$$\sup_{y \in \Omega} \frac{\bar{L}_\lambda \psi}{\psi} \geq \mu(\lambda). \tag{5.12}$$

Since $\bar{L}_\lambda \phi = \mu(\lambda)\phi$, the formula (5.10) follows. Note that we do not require the test functions ψ to be $C^2(\Omega)$, only $\Delta \psi \in C(\Omega)$. This is important, since we do not want to require the shear $b(y)$ to be any more regular than $b \in C(\bar{\Omega})$.

Let us derive upper and lower bounds for $\mu(\lambda)$ by choosing test functions ψ as

$$\psi = 1 + \lambda \delta \chi + \lambda^2 \delta^2 h, \tag{5.13}$$

where $\chi = \chi(y)$ and $h = h(y)$ solve

$$\Delta \chi = -b, \quad \Delta h = -b\chi + k, \tag{5.14}$$

with zero Neumann boundary conditions at $\partial \Omega$, and k a constant equal to

$$k = \frac{1}{|\Omega|} \int_\Omega b\chi \, dy = \frac{1}{|\Omega|} \int_\Omega |\nabla \chi|^2 \, dy. \tag{5.15}$$

We normalize χ and h so that

$$\inf_{x \in \Omega} \chi(x) = 0, \quad \inf_{x \in \Omega} h(x) = 0. \tag{5.16}$$

Then

$$\bar{L}_\lambda \psi = \lambda^2 \delta^2 k + \lambda^3 \delta^3 bh + (\lambda^2 + f'(0))\psi$$

and

$$\frac{(\bar{L}_\lambda \psi, \psi)}{\|\psi\|_2^2} = \lambda^2 \delta^2 k \frac{\int \psi}{\int \psi^2} + \lambda^3 \delta^3 \frac{\int bh\psi}{\int \psi^2} + \lambda^2 + f'(0). \tag{5.17}$$

Using the definition of ψ, we see that $\psi = \psi^2 - \lambda \delta \chi \psi - \lambda^2 \delta^2 h \psi$ and

$$\frac{\int_\Omega \psi \, dy}{\int_\Omega \psi^2 \, dy} = 1 - \lambda \delta \frac{\int_\Omega \chi \psi \, dy}{\int_\Omega \psi^2 \, dy} - \lambda^2 \delta^2 \frac{\int_\Omega h \psi \, dy}{\int_\Omega \psi^2 \, dy}.$$

Now from (5.9) and (5.17) we have the lower bound

$$\mu(\lambda) \geq \lambda^2 + f'(0) + \lambda^2 \delta^2 k + R_1, \tag{5.18}$$

with

$$R_1 = -\frac{\lambda^3 \delta^3}{\int_\Omega \psi^2} \left(k \int_\Omega \chi \psi + k \lambda \delta \int_\Omega h \psi - \int_\Omega b h \psi \right). \tag{5.19}$$

By choice of $\chi \geq 0$ and $h \geq 0$, we have $\int_\Omega \psi^2 \geq |\Omega|$ for all $\delta \geq 0$ and $\lambda > 0$. Hence, $R_1 = O(\delta^3)$ for λ bounded. Returning to the variational formula (5.8), we now have a lower bound on $c^*(\delta)$:

$$c^*(\delta) = \inf_{0 < \lambda \leq \lambda_0} \frac{\mu(\lambda)}{\lambda} \geq \inf_{0 < \lambda \leq \lambda_0} \left(\lambda + \frac{f'(0)}{\lambda} + \lambda \delta^2 k + \frac{R_1}{\lambda} \right)$$

$$\geq \inf_{\lambda > 0} \left(\lambda + \frac{f'(0)}{\lambda} + \lambda \delta^2 k \right) + O(\delta^3)$$

$$= 2\sqrt{f'(0)(1 + \delta^2 k)} + O(\delta^3)$$

$$= c_0 + \frac{c_0 \delta^2 k}{2} + O(\delta^3). \tag{5.20}$$

To obtain an upper bound on $c^*(\delta)$, we use (5.10) and calculate

$$\frac{\bar{L}_\lambda \psi}{\psi} = \frac{\Delta \psi}{\psi} + \lambda \delta b + \lambda^2 + f'(0) = \frac{\lambda^2 \delta^2 k + \lambda^3 \delta^3 b h}{1 + \lambda \delta \chi + \lambda^2 \delta^2 h} + \lambda^2 + f'(0). \tag{5.21}$$

Since $\chi \geq 0$ and $h \geq 0$, we see from (5.10) and (5.21) that

$$\mu(\lambda) \leq \sup_{y \in \Omega} \frac{\bar{L}_\lambda \psi}{\psi} \leq \lambda^2 + f'(0) + \lambda^2 \delta^2 k + R_2, \tag{5.22}$$

with

$$R_2 = \lambda^3 \delta^3 \|b h\|_\infty. \tag{5.23}$$

The variational formula (5.8) implies

$$c^*(\delta) = \inf_{0 < \lambda \leq \lambda_0} \frac{\mu(\lambda)}{\lambda} \leq \inf_{0 < \lambda \leq \lambda_0} \left(\lambda + \frac{f'(0)}{\lambda} + \lambda \delta^2 k + R_2 \right)$$

$$= c_0 + \frac{c_0 \delta^2 k}{2} + O(\delta^3), \tag{5.24}$$

thus completing the proof. $\square$

When the shear $b(y, \omega)$ is a random process, the corresponding minimal speed $c^*(\delta) = c^*(\delta, \omega)$ is a random variable for each δ, and we consider how the expectation $E[c^*(\delta)]$ scales with the parameter δ by finding an exponent p such that $E[c^*(\delta)] = c^*(0) + O(\delta^p)$. Each realization of the random process $b(y, \omega)$ restricted to the domain Ω does not necessarily have zero integral over Ω. Nevertheless, each realization can be written in the form

$$b(y, \omega) = \bar{b}(\omega) + b_1(y, \omega), \tag{5.25}$$

where $\bar{b}(\omega) = \langle b(y, \omega) \rangle$ is the mean of b over Ω, and $b_1(y, \omega)$ is the variation about the mean value. For a fixed realization, the minimal speed $c^*(\delta)$ will be affected by both the scaling of the mean $\bar{b}(\omega)$ and the scaling of the variation $b_1(y, \omega)$. That is,

$$c^*(\delta, \omega) = c_0^* + \delta \bar{b}(\omega) + M(\delta, \omega), \tag{5.26}$$

where the remainder $M(\delta, \omega)$ is the enhancement due to the variation $b_1(y, \omega)$, different for each realization. Taking the expectation of both sides of (5.26), we have

$$E[c^*(\delta)] = c_0^* + \delta E[\bar{b}(\omega)] + E[M(\delta, \omega)]. \tag{5.27}$$

Though for each sample, $M(\delta, \omega)$ is $O(\delta^2)$ for δ small, we show that $E[M(\delta)]$ exhibits the same quadratic scaling for enhancement of averaged front speeds under suitable moment conditions of the shear.

Theorem 5.3. *Let $b(y, \omega)$ be a stationary random process in $\mathbb{R}^{n-1}$ ($n \geq 2$) such that sample paths are almost surely continuous and such that*

$$E[\|b\|_\infty^6] < +\infty. \tag{5.28}$$

Then for δ small, the expectation $E[c^*(\delta)]$ has the expansion

$$E[c^*(\delta)] = c_0 + \delta E[\langle b \rangle] + \frac{c_0 \delta^2}{2|\Omega|} \int_\Omega E[|\nabla \chi|^2] \, dy + O(\delta^3), \tag{5.29}$$

where $b(y, \omega) = \langle b \rangle(\omega) + b_1(y, \omega)$ and $\chi = \chi(y, \omega)$ solves $\Delta_y \chi = -b_1$, $y \in \Omega$, subject to zero Neumann boundary condition.

Proof. Since the contribution of the integral average $\langle b \rangle$ to c^* is an additive constant, it suffices to consider shear flow b_1 and show that it gives the averaged speed

$$E[c^*(\delta)] = c_0 + \frac{c_0 \delta^2}{2|\Omega|} \int_\Omega E[|\nabla \chi|^2] \, dy + O(\delta^3). \tag{5.30}$$

We adapt the proof of Proposition 5.1, noting that in the stochastic case the remainders R_1 and R_2 defined by (5.19) and (5.23) are random and not bounded uniformly for all realizations. Instead, we will show that for λ in a bounded interval,

$$E[|R_1|] \leq O(\delta^3), \quad E[|R_2|] \leq O(\delta^3).$$

To this end, we estimate χ and h, with C denoting a generic positive constant depending only on the domain Ω and its dimension. Let χ and h solve (5.14) with zero integral averages $\langle \chi \rangle = \langle h \rangle = 0$. Applying $W^{2,p}$ estimates [102], we have

$$\|\chi\|_{W^{2,p}(\Omega)} \leq C\|b_1\|_{L^p(\Omega)} \leq C|\Omega|^{1/p}\|b_1\|_\infty \tag{5.31}$$

and

$$\begin{aligned}
\|h\|_{W^{2,p}(\Omega)} &\leq C\|b_1\chi + k\|_{L^p(\Omega)} \\
&\leq C\|b_1\|_\infty\|\chi\|_{L^p(\Omega)} + Ck|\Omega|^{1/p} \\
&\leq C\|b_1\|_\infty^2,
\end{aligned} \tag{5.32}$$

since

$$k = \langle |\nabla \chi|^2 \rangle \leq C\|b_1\|_\infty^2. \tag{5.33}$$

Given $\alpha \in (0,1)$, we can choose $p > 1$ sufficiently large, depending on n, such that $W^{2,p}(\Omega)$ embeds continuously into $C^{1,\alpha}(\bar{\Omega})$. It follows that there is a constant $C > 0$ independent of b such that

$$\|\chi\|_{C^1(\bar{\Omega})} \leq C\|b_1\|_\infty, \quad \|h\|_{C^1(\bar{\Omega})} \leq C\|b_1\|_\infty^2. \tag{5.34}$$

If instead we normalize χ and h by (5.16), then the bounds (5.34) still hold, with different constants. Note that adding a constant to χ and h does not alter the quantity $\int_\Omega |\nabla\chi|^2$ that appears in the asymptotic expansion.

Now by (5.34), the integrals in R_1 are easily bounded as

$$\begin{aligned}
\int_\Omega \chi\psi &= \int_\Omega \chi + \lambda\delta\chi^2 + \lambda^2\delta^2\chi h \\
&\leq C\left(\|b_1\|_\infty + \lambda\delta\|b_1\|_\infty^2 + \lambda^2\delta^2\|b_1\|_\infty^3\right).
\end{aligned}$$

Similarly,

$$\begin{aligned}
\int_\Omega h\psi &= \int_\Omega h + \lambda\delta\chi h + \lambda^2\delta^2 h^2 \\
&\leq C\left(\|b_1\|_\infty^2 + \lambda\delta\|b_1\|_\infty^3 + \lambda^2\delta^2\|b_1\|_\infty^4\right)
\end{aligned}$$

and

$$\begin{aligned}
\left|\int_\Omega b_1 h\psi\, dy\right| &= \left|\int_\Omega b_1 h + \lambda\delta\int_\Omega b_1 h\chi + \lambda^2\delta^2\int_\Omega b_1 h^2\right| \\
&\leq C\left(\|b_1\|_\infty^3 + \lambda\delta\|b_1\|_\infty^4 + \lambda^2\delta^2\|b_1\|_\infty^5\right).
\end{aligned}$$

Since χ and h are nonnegative, it follows that $\int_\Omega \psi^2\, dy \geq C\int_\Omega \psi = C|\Omega| > 0$, for any realization. So for λ in a bounded interval and δ small, we bound (5.19) by

$$|R_1| \leq C\delta^3\lambda^3(1 + \|b_1\|_\infty^6),$$

so that $E[|R_1|] \leq O(\delta^3)$.

To bound $E[|R_2|]$, we use the normalization (5.16) and the above estimates:

$$|R_2| = \lambda^3 \delta^3 \|b_1 h\|_\infty \leq C\lambda^3 \delta^3 \|b_1\|_\infty^3. \tag{5.35}$$

Hence $E[R_2] \leq O(\delta^3)$ for λ in a finite interval. Now we return to (5.20) to conclude that

$$E[c^*(\delta)] \geq E[2\sqrt{f'(0)(1+\delta^2 k)}] + O(\delta^3) = c_0 + \frac{c_0 \delta^2 E[k]}{2} + O(\delta^3),$$

since $E[k^2] \leq CE[\|b\|_\infty^4] < \infty$.

The opposite inequality follows from (5.24), since $E[R_2] = O(\delta^3)$ for $\lambda \in (0, \lambda_0)$. Thus formula (5.30) holds. For general b, not necessarily mean zero, we have

$$E[c^*(\delta)] = c_0 + \delta E[\langle b \rangle] + \frac{c_0 \delta^2}{2} E[k] + O(\delta^3)$$

$$= c_0 + \delta E[\langle b \rangle] + \frac{c_0 \delta^2}{2|\Omega|} \int_\Omega E[|\nabla \chi|^2] \, dy + O(\delta^3).$$

The proof is finished.   □

If we choose the Ornstein–Uhlenbeck (O-U) process in Chapter 1 for shear b, then $E[\langle b \rangle] = 0$. Let us show below that the O-U process, denoted by $X(y, \omega)$, satisfies the conditions in Theorem 5.3, and so $E[c^*(\delta)]$ scales quadratically with δ for δ small.

Corollary 5.4 (Explicit Average Speed Formula).

Consider the O-U process $b(y, \omega)$ as a solution of the stochastic differential (Itō) equation

$$dX(y) = -aX(y)\,dy + r\,dW(y), \quad y \in [0, L], \tag{5.36}$$

where $W(y, \omega)$ is the standard Wiener process, and $X(0, \omega) = X_0(\omega)$ is a Gaussian random variable with mean zero and variance $\rho = r^2/(2a)$. Then $X(y, \omega)$ satisfies the moment conditions in Theorem 5.3. The averaged KPP front speed in the channel $\mathbb{R} \times [0, L]$ is given by

$$E[c^*(\delta)] = c_0 + \frac{c_0 \delta^2}{2} \text{enh} + O(\delta^3), \quad \delta \ll 1, \tag{5.37}$$

where

$$\text{enh} = \frac{r^2}{2a} \left(e^{-aL} \left(\frac{4}{L^2 a^4} - \frac{1}{3a^2} \right) + \frac{L}{3a} - \frac{4}{L^2 a^4} - \frac{5}{3a^2} + \frac{4}{La^3} \right).$$

Proof. The O-U process is stationary and Markov. Its sample paths are almost surely Hölder continuous though nowhere differentiable. The process can be written as

$$b(y,\omega) = e^{-ay}b(0,\omega) + r \int_0^y e^{-a(y-s)} dW_s(\omega). \tag{5.38}$$

The covariance function of this process is $\rho e^{-a|y-s|}$. Letting $g(y,\omega)$ denote the process

$$g(y) = e^{ay}b(y,\omega) = g(0,\omega) + r \int_0^y e^{as} dW_s(\omega), \tag{5.39}$$

we see that $g(y,\omega)$ is a martingale [125]. By Doob's martingale moment inequality [125], for any $p \in (1,+\infty)$,

$$E\left[\sup_{0<y<L} |g(y)|^p\right] \leq \left(\frac{p}{p-1}\right)^p E[|g(L)|^p]. \tag{5.40}$$

Since the process $b(y,\omega)$ is Gaussian, (5.39) and (5.40) imply that

$$E\left[\|b\|_\infty^6\right] \leq CE[|b(L)|^6] < +\infty. \tag{5.41}$$

Formula (5.37) now applies to the average speed. Notice that

$$(\chi_x(x))^2 = \int_0^x \int_0^x b_1(s)b_1(y)\,ds\,dy$$

and

$$E[(\chi_x(x))^2] = \int_0^x \int_0^x E[b_1(s)b_1(y)]\,ds\,dy. \tag{5.42}$$

Let us calculate $E[b_1(s)b_1(y)]$ in terms of $E[b(s)b(y)]$. Define

$$g(y) = \langle f(\cdot,y)\rangle, \quad \text{or} \quad g(s) = \langle f(s,\cdot)\rangle,$$

so that

$$E[b_1(y)b_1(s)] = E[b(s)b(y)] - E[b(s)\bar{b}] - E[b(y)\bar{b}] + E[\bar{b}^2],$$
$$E[b(s)\bar{b}] = \frac{1}{L}\int_0^L E[b(y)b(s)]\,dy = g(s),$$
$$E[\bar{b}^2] = \frac{1}{L^2}\int_0^L \int_0^L E[b(s)b(y)]\,dy\,ds = \langle g\rangle.$$

Thus $E[b_1(y)b_1(s)] = f(s,y) + \langle g\rangle - g(y) - g(s)$. Now, we have

$$E[(\chi_x(x))^2] = \int_0^x \int_0^x E[b_1(s)b_1(y)]\,ds\,dy$$
$$= \int_0^x \int_0^x f(s,y) + \langle g\rangle - g(y) - g(s)\,ds\,dy$$
$$= x^2\langle g\rangle - 2x^2\langle g\rangle_x + \int_0^x \int_0^x f(s,y)\,ds\,dy,$$

where $\langle g\rangle_x$ denotes the average of g over the interval $[0,x]$, for $0 < x \le L$. Consequently, we have

$$E[\langle|\chi_x|^2\rangle] = \frac{1}{L}\int_0^L \left(x^2\langle g\rangle - 2x^2\langle g\rangle_x + \int_0^x\int_0^x f(s,y)\,ds\,dy\right)dx. \qquad (5.43)$$

Using the O-U covariance function, we obtain

$$g(s) = \frac{1}{L}\int_0^L f(s,y)\,dy = \frac{r^2}{2a}\frac{1}{L}\int_0^L e^{-a|y-s|}\,dy$$
$$= \frac{r^2}{2a}\left(\frac{1-e^{-as}}{La} + \frac{1-e^{-a(L-s)}}{La}\right)$$

and

$$\langle g\rangle_x = \frac{r^2}{2a}\frac{1}{x}\int_0^x\left(\frac{1-e^{-as}}{La} + \frac{1-e^{-a(L-s)}}{La}\right)ds$$
$$= \frac{r^2}{2a}\left(\frac{2}{La} + \frac{1}{xLa^2}\left(e^{-ax}-1\right) + \frac{1}{xLa^2}\left(e^{-aL}-e^{-a(L-x)}\right)\right).$$

Letting $x = L$, we have

$$\langle g\rangle = \frac{r^2}{2a}\left(\frac{2}{La} + \frac{2}{L^2a^2}\left(e^{-aL}-1\right)\right).$$

Similarly,

$$\int_0^x\int_0^x f(s,y)\,ds\,dy = \frac{r^2}{2a}\left(\frac{2x}{a} + \frac{2}{a^2}\left(e^{-ax}-1\right)\right).$$

Combining the above, we have

$$E[\langle|\chi_x|^2\rangle] = \frac{r^2}{2a}\left(\frac{2L}{3a} + \frac{2}{3a^2}\left(e^{-aL}-1\right)\right)$$
$$- \frac{r^2}{2a}\frac{2}{L}\int_0^L \frac{2x^2}{La} + \frac{x}{La^2}\left(e^{-ax}-1\right) + \frac{x}{La^2}\left(e^{-aL}-e^{-a(L-x)}\right)dx$$
$$+ \frac{r^2}{2a}\frac{1}{L}\int_0^L \frac{2x}{a} + \frac{2}{a^2}\left(e^{-ax}-1\right)dx$$
$$= \frac{r^2}{2a}\left(e^{-aL}\left(\frac{4}{L^2a^4} - \frac{1}{3a^2}\right) + \frac{L}{3a} - \frac{4}{L^2a^4} - \frac{5}{3a^2} + \frac{4}{La^3}\right). \qquad (5.44)$$

In view of (5.37), the proof is complete.   □

We see from formula (5.37) that the correction term of $E[c^*]$ diverges as the domain width L goes to $+\infty$, suggesting that KPP front speed in a Gaussian spatial shear flow on the whole plane $\mathbb{R}^2$ is infinite.

Theorem 5.5 (Linear Growth). *If the stationary shear process $b(y,\omega)$ has almost surely continuous sample paths and satisfies $E[\|b\|_\infty] < \infty$, then the amplified shear field $\delta b(y,\omega)$ generates the average front speed:*

$$E[|c^*(\delta,\omega)|] = O(\delta), \quad \delta \gg 1.$$

Moreover, $\lim_{\delta \to \infty} E[|c^*(\delta,\omega)|]/\delta$ exists.

Proof. By (2.80), for all ω, $|c*(\delta,\omega)|/\delta \to d^*(\omega)$ as $\delta \to \infty$, where d^* is finite for each ω. Now recall the upper bound $|c^*(\delta,\omega)| \leq |c_0| + \delta\|b\|_\infty$. Hence for $\delta > |c_0|$, we have $|c^*(\delta,\omega)|/\delta \leq 1 + \|b\|_\infty \equiv Y$, and $E(Y) < \infty$. The dominated convergence theorem implies that

$$E\left[\frac{|c*(\delta,\omega)|}{\delta}\right] \to E[d^*(\omega)] \leq E(Y).$$

The proof is finished.   □

The O-U process satisfies the required condition for linear average speed growth. By (2.80), the growth rate $d^*(\omega)$ is

$$d^*(\omega) = \sup_{\psi \in D_1} \int_\Omega b(y,\omega)\psi^2(y)\,dy,$$

where

$$D_1 = \left\{\psi \in H^1(\Omega) : \|\nabla\psi\|_2^2 \leq f'(0), \|\psi\|_2 = 1\right\}.$$

If a realization of b were to have a flat piece near the maximal point of b in Ω, then choosing the test function ψ supported near the maximal point within the flat piece would imply that $d^*(\omega)$ equals $\sup_\Omega b(y,\omega)$.

However, this happens with zero probability for an O-U process because it is oscillatory in every small neighborhood of the maximal point. The distribution of $d^*(\omega)$ appears to be an open problem. However, numerical simulation suggests in certain parameter regimes (next subsection) that $d^*(\omega)$ and $\sup_\Omega b(y,\omega)$ are quite correlated. The latter is a running maximum and has a well-studied limit law as the domain size grows; see Theorem 4.9.

5.1.2 Computing Front Speeds by the Variational Principle

Let us briefly discuss the numerical procedure of computing $c^*(\omega)$ and its statistics. Let $n = 2$. For a given $\lambda > 0$, we compute the principal eigenvalue $\mu(\lambda)$ with corresponding eigenfunction $\phi = \phi(y) > 0$, $y \in [0, L]$, by solving

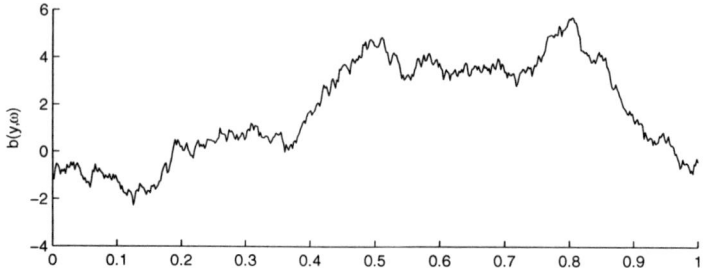

Figure 5.2 One sample path of the Ornstein–Uhlenbeck process $b(y, \omega)$, from [178].

$$\phi_{yy} + [\lambda^2 + \lambda\, b(y) + f'(0)]\phi = \mu(\lambda)\phi, \quad y \in (0, L),$$

$$\frac{\partial \phi}{\partial y} = 0, \quad y = 0, L, \tag{5.45}$$

using a standard second-order finite difference method. Here we suppress the random parameter ω, since computation is done realization by realization. Denote the uniform partition of the domain by points by $\{y_i\}_{i=1}^{m}$ and the numerical solution by $\bar{\phi} = \{\bar{\phi}_i\}_{i=1}^{m}$, where $h = L/(m-1)$, $y_i = (i-1)h$, and $\bar{\phi}_i \approx \phi(y_i)$. The discretized system is

$$\frac{1}{h^2}\bar{\phi}_{i-1} + \left(\lambda^2 + \lambda b_i + f'(0) - \frac{2}{h^2}\right)\bar{\phi}_i + \frac{1}{h^2}\bar{\phi}_{i+1} = \mu(\lambda)\bar{\phi}_i \quad i = 2, \ldots, m-1,$$

with second-order approximation of the Neumann boundary conditions. This reduces to finding the principal eigenvalue of a symmetric tridiagonal matrix, a problem with efficient algorithms in numerical linear algebra such as double-precision LAPACK routines [6]. Then we compute points on the curve $H(\lambda) = \mu(\lambda)/\lambda$, and minimize over λ using a Newton's method with line search.

Our approximation decreases H with each iteration and converges quadratically in the region near the infimum. We generate realizations of the shear process $b(y, \omega)$ by numerically evaluating the stochastic ODE (5.36) by a second-order-accurate scheme in the parameter h [133]. Figure 5.2 shows an O-U sample path. The O-U parameters are $\rho = 2, a = 4$.

To approximate the expectation $E[c^*(\delta)]$, we generate N independent realizations (indexed by $i = 1, \ldots, N$) of the shear and compute the corresponding minimal speeds $\{c_i^*\}$ for each δ. Then we compute the average

$$E[c^*(\delta)] \approx \bar{E}(\delta) = c_0^* + \frac{1}{N}\sum_{i=1}^{N} M_i(\delta), \tag{5.46}$$

where $M_i(\delta) = c_i^*(\delta) - c_0^* - \delta\bar{b}_i$. That is, we subtract the linear part due to the mean of the shear being nonzero, as in (5.26). Once we have the averages $\bar{E}(\delta)$ for each δ,

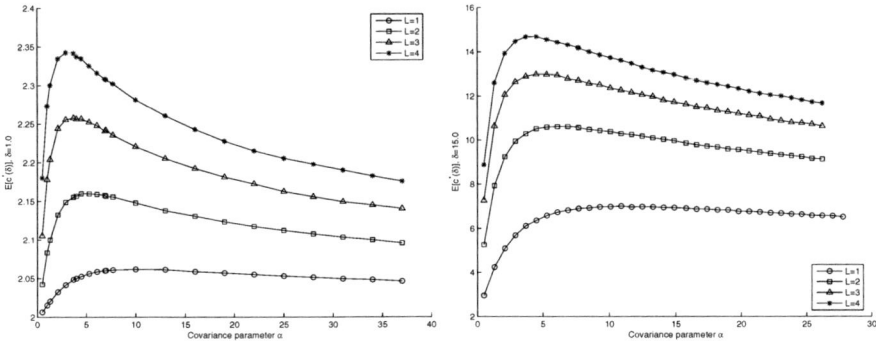

Figure 5.3 Effect of covariance parameter α on the speed enhancement for $L = 1$ (line–circle), 2 (line–square), 3 (line–triangle), 4 (line–star) at $\delta = 1.0$ (left) and $\delta = 15.0$ (right), from [178].

we compute the exponents p using the least-squares method to fit a line to a log-log plot of speed versus amplitude. That is, the exponent p is the slope of the best-fit line through the data points $(\log(\delta), \log(\bar{E}[c^*(\delta)] - c^*(0)))$ for each shear amplitude δ.

Besides confirming the quadratic and linear laws of $E[c^*]$, we perform a parameter study of the dependence of the statistics of c^* on the correlation length of the flow and domain width. Let us consider the effect of the covariance on the enhancement of c^*. The covariance $E[b(y)b(s)]$ is a function of $|t| = |s - y|$, so we write $V(t) = E[b(y)b(s)]$. By choosing $r = \sqrt{2}\alpha^{3/4}$, we construct O-U processes with covariances given by

$$V(t) = \sqrt{\alpha}e^{-\alpha|t|} \tag{5.47}$$

By this choice of r, the L^2 norm of $V(t)$ remains constant as α changes, so that the total energy in the power spectrum of the signal remains constant. Since $\frac{r^2}{2a} = \sqrt{a}$, we see from equation (5.37) that for fixed L,

$$\lim_{\alpha \to +\infty} E[\langle|\chi_x|^2\rangle] = \lim_{\alpha \to 0^+} E[\langle|\chi_x|^2\rangle] = 0, \tag{5.48}$$

and that $E[\langle|\chi_x|^2\rangle]$ achieves a maximum for some finite value of $\alpha \in (0, \infty)$. This suggests that there is some optimal α, depending on the domain size L, such that the enhancement of $E[c^*(\delta)]$ is maximized.

Fixing the grid spacing $dx = 0.002$, we computed the expected value $E[c^*(\delta)]$ for a range of α and for $L = 1.0, 2.0, 3.0, 4.0$. Note that for each α, we must choose the initial points b_0 to have variance $E[b_0^2] = \sqrt{\alpha}$, so that the process remains stationary for each α. Varying the covariance does not affect the order of the scaling in δ. That is, in each case the enhancement scales like $O(\delta^2)$ for small δ and $O(\delta)$ for large δ.

Figure 5.3 shows the enhancement $E[c^*(\delta)]$ for $\delta = 1, 15$, and a range of α. The enhancement peaks at an optimal covariance parameter α.

This resonance effect can be interpreted in terms of $V(t)$ and its Fourier transform (power spectrum)

$$\hat{V}(w) = \sqrt{\frac{2}{\pi\alpha}} \left(1 + \left(\frac{w}{\alpha}\right)^2\right)^{-1}. \tag{5.49}$$

As $\alpha \to 0$, $\hat{V}(w)$ concentrates at the origin, and so the energy of the shear process is concentrated more in the large-scale spatial modes. The domain Ω, to which the process is restricted is bounded, and variations over a length scale that is much greater than the diameter of Ω have little effect on the average enhancement of the front. As a result, $E[c^*]$ decreases as $\alpha \to 0^+$. In the other limit $\alpha \to \infty$, $\hat{V}$ spreads out so that the energy over any finite band of frequencies goes to zero, causing $E[c^*]$ to decrease as well. Note that $V \to 0$ in L^1 as $\alpha \to \infty$, so even though $\hat{V}$ spreads out more uniformly as $\alpha \to 0$, the family of processes does not converge to white noise, whose covariance function equals the Dirac delta function.

We also observe that as L increases with δ fixed, the expectation $E[c^*(\delta)]$ grows sublinearly and the variance $\mathrm{Var}[c^*(\delta)]$ decreases. The average speed obeys the upper bound

$$E[c^*(\delta)] \le c_0^* + \delta E[\sup_{y \in [0,L]} b(y)]. \tag{5.50}$$

We modify the diffusion constant to equal to 0.01 (set to 1 in equation (5.1)). Recall from (2.80) that in the regime of small diffusion coefficient, the ratio $c^*(\delta)/\delta$ is close to $\sup_{y \in [0,L]} b(y)$ for $\delta \gg 1$. By Theorem 4.9 and (5.50), the growth of $E[c^*(\delta)]$ with respect to L can be no more than $O(\sqrt{\log L})$.

We observed that at $\delta = 50$, diffusion constant is 0.01, and the mean and variance of $c^*(\delta)/\delta$ mimic the mean and variance of $\sup_{y \in [0,L]} b(y)$ as $L \gg 1$. In Figure 5.4, we compare $E[c^*(\delta)]$ with $E[g_1(L)]$ and $\mathrm{Var}[c^*(\delta)]$ with $\mathrm{Var}[g_1(L)]$, where

$$g_1(L) = c_0^* + \delta \sup_{y \in [0,L]} b(y).$$

The figure shows a close correlation between the speed and the maximum of the shear on $[0,L]$, though the curves are clearly not identical. The tracking of c^* by the running maximum of the shear in this regime is another indication of the divergence of c^* as $L \to +\infty$. This is very similar to the divergence of homogenization of HJ equations with classical Hamiltonians and O-U potentials in Chapter 4. An explanation and another perspective will be presented in the next section.

Speed Distribution

For a fixed $\delta = 1$ and $\delta = 14$ (corresponding to small and large amplitudes), we computed the distributions of the numerically computed values $M(\delta)$. To compute these distributions, we partition the range of values into Q disjoint intervals: $\{[x_j, x_{j+1})\}_{j=1}^{Q}$. Then we let

$$\mathrm{pdf}(x) = \frac{1}{N} \sum_{i=1}^{N} \frac{\chi_j(M_i(\delta))}{(x_{j+1} - x_j)} \quad \text{if } x \in [x_j, x_{j+1}), \tag{5.51}$$

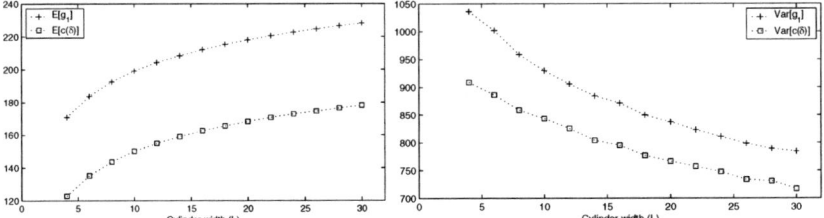

Figure 5.4 Sublinear growth of ensemble-averaged speed (dot–square) as domain width increases (left); decay of the speed variance (dot–square) as domain width increases (right); $\delta = 50$. Both are tracked by mean and variance of the upper bound g_1 (dot–plus) in (5.50); from [175].

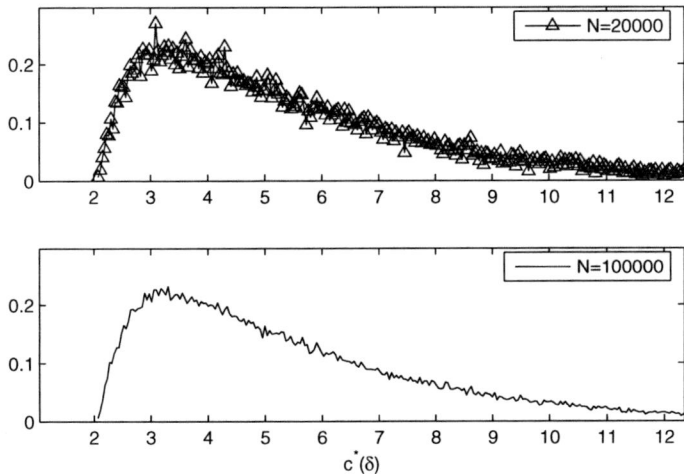

Figure 5.5 Convergence of speed-enhancement distribution with increasing number of random samples at $\delta = 14.0$; from [178].

where $\chi_j(x)$ is the characteristic function of the interval $[x_j, x_{j+1})$. The values $M(\delta)$ are the enhancement of the minimal speeds due to the variation of the shear after the effect of the mean field has been subtracted off. Since a mean-zero shear always enhances the minimal speeds, we expect $M(\delta) > 0$ for all δ, for all realizations. Figure 5.5 shows the convergence of the distributions at $N = 100,000$ samples, and $Q = 300$.

The computation of KPP front speeds in random shear flows inside three-dimensional cylinders ($D = \mathbb{R} \times \Omega$, with $\Omega \subset \mathbb{R}^2$ a bounded cross section with varying shapes) has recently been performed [215]. The principal eigenvalue μ in a general-shaped domain Ω is computed by a two-scale finite element method. It is found that if the area of Ω is kept invariant, then the larger the domain aspect ratio (narrower cross section), the larger the ensemble-averaged front speeds.

5.2 KPP Fronts in Temporally Random Shear Flows

In this section, we consider KPP fronts in time-random shear flows and derive the corresponding variational formula of c^* based on [179, 247]. In previous formulas, c^* is associated with the principal eigenvalue of an elliptic operator in case of spatial heterogeneities or a periodic-parabolic operator in case of media with time-periodic variations. In time-random media, the principal Lyapunov exponent replaces the principal eigenvalue and is almost surely deterministic if there is sufficient mixing in the random media. We shall explore the connection of c^* with a well-studied topic called the parabolic Anderson problem in probability, and the homogenization of the viscous version of the stochastic Hamilton–Jacobi equations in Chapter4.

5.2.1 Lyapunov Exponent, Large Deviation, and Front Spreading

The KPP equation is

$$u_t = \frac{1}{2}\Delta_z u + B \cdot \nabla_z u + f(u), \tag{5.52}$$

where $u = u(z,t)$, $z = (x,y) \in \mathbb{R} \times \mathbb{R}^{n-1}$, $n \geq 2$; f is KPP, with compactly supported initial data bounded between 0 and 1; and $B = (b(y,t),0,\ldots,0)$, $b(y,t)$ is a stationary Gaussian process in t, with a deterministic profile in y. More precisely, the function $b(y,t) = b(y,t,\hat{\omega})$ is a mean-zero Gaussian random field over (y,t), periodic in y with period L for each fixed t, and stationary in t for each fixed y. The field b is defined over the probability space $(\hat{\Omega}, \hat{\mathscr{F}}, Q)$ and has covariance function $\Gamma(y_1,y_2,t_1,t_2) = E_Q[b(y_1,t_1)b(y_2,t_2)]$. The following assumptions hold on $b(y,t)$:

A1: (Periodicity in y) Let $C_P^{0,1}(D)$ denote the space of Lipschitz continuous functions that are periodic on the period cell $D = [0,L]^{n-1}$. For each $\hat{\omega} \in \hat{\Omega}$, there is a continuous map $J(\cdot,\hat{\omega}) : [0,+\infty) \to C_P^{0,1}(D)$ such that $b(\cdot,t,\hat{\omega}) = J(t,\hat{\omega})$.

A2: (Stationarity in t) For each $s \in \mathbb{R}^+$ there is a measure-preserving transformation $\tau_s : \hat{\Omega} \to \hat{\Omega}$ such that $b(y,\cdot+s,\hat{\omega}) = b(y,\cdot,\tau_s\hat{\omega})$. Hence, Γ depends only on y_1,y_2 and $|t-s|$.

A3: (Ergodicity) The transformation τ_s is ergodic: if a set $A \in \hat{\mathscr{F}}$ is invariant under the transformation τ_s, then either $Q(A) = 0$ or $Q(A) = 1$.

A4: The field b has mean zero, is almost surely continuous in (y,t), and has uniformly bounded variance:

$$E[b(y,t)] = 0, \quad E\left[b(y,t)^2\right] \leq \sigma^2 \ \forall y \in D, \ t \geq 0. \tag{5.53}$$

A5: (Decay of temporal correlations) The function $\hat{\Gamma}(r) = \sup_{y_1,y_2}\Gamma(y_1,y_2,0,r)$ is integrable over $[0,\infty)$:

$$\int_0^\infty \hat{\Gamma}(r)\,dr = p_1 < \infty \qquad (5.54)$$

for some finite constant $p_1 > 0$. This constant will appear later in estimates of the front speed.

A6: There is a finite constant $p_2 > 0$ such that

$$|\Gamma(y_1, y_2, s, t) - \Gamma(y_1, y_3, s, t)| \leq p_2 |y_3 - y_2| \hat{\Gamma}(|s - t|).$$

For example, a field satisfying assumptions A1–A6 might have the form

$$b(y, t, \hat{\omega}) = \sum_{j=1}^N b_1^j(y) b_2^j(t, \hat{\omega}),$$

where the functions $b_1^j(y)$ are deterministic, Lipschitz continuous, and periodic over D, and the functions $b_2^j(t, \hat{\omega})$ are mean-zero stationary Gaussian fields in t.

Before stating the main results, let us define the family of Markov processes associated with the linear part of the operator in (5.52). For a fixed $\hat{\omega} \in \hat{\Omega}$ and for each $z \in \mathbb{R}^n$, $t \geq 0$, let $Z^{z,t}(s) = (X^{z,t}(s), Y^{z,t}(s)) \in \mathbb{R}^n$ solve the Itô equation

$$dZ^{z,t}(s) = B(Z^{z,t}(s), t - s)\,ds + dW(s), \quad s \in [0, t], \qquad (5.55)$$

with initial condition $Z^{z,t}(0) = z = (x, y) \in \mathbb{R}^n$, where $W(s) = (W_1(s), W_2(s)) \in \mathbb{R}^n$ is the n-dimensional Wiener process with $W(0) = 0$. Because of the shear structure of B, we therefore have

$$X^{z,t}(s) = x + \int_0^s b(y + W_2(\tau), t - \tau)\,d\tau + W_1(s), \qquad (5.56)$$

$$Y^{z,t}(s) = y + W_2(s).$$

Let $P^{z,t}$ denote the corresponding family of measures on $C([0, t]; \mathbb{R}^n)$. These stochastic trajectories will play the role of characteristics in hyperbolic problems or candidates for minimizing paths in the Lax formula of HJ analysis.

First, let us define the analogue of the principal eigenvalue in the KPP analysis of Chapter 2 (see (2.51)–(2.52), (2.76)):

Proposition 5.6. *Assume that A1–A6 hold for the process $b(y, t)$. There is a set $\hat{\Omega}_0 \subset \hat{\Omega}$ such that $Q(\hat{\Omega}_0) = 1$ and for any $\hat{\omega} \in \hat{\Omega}_0$ and any $\lambda \in \mathbb{R}^n$, the limit*

$$\mu(\lambda, z) = \mu(\lambda) = \lim_{t \to \infty} \frac{1}{t} \log E\left[e^{-\lambda \cdot (Z^{z,t}(t) - z)} \right] \qquad (5.57)$$

exists uniformly over $z \in \mathbb{R}^n$ and locally uniformly over $\lambda \in \mathbb{R}^n$. The limit $\mu(\lambda)$ is a finite constant for all $\hat{\omega} \in \hat{\Omega}$ and is independent of $z \in \mathbb{R}^n$. Moreover, $\mu(\lambda) \geq 0$, and $\mu(\lambda)$ is both convex and superlinear: $\mu(\lambda)/|\lambda| \to +\infty$ as $|\lambda| \to \infty$.

The convexity and coercivity of μ is reminiscent of basic assumptions in the HJ homogenization of Chapter 4. The Legendre transform of $\mu(\lambda)$ shares the same property and equals

$$S(c) = \sup_{\lambda \in \mathbb{R}^n} [c \cdot \lambda - \mu(\lambda)]. \tag{5.58}$$

In the language of classical mechanics, the function μ will play the role of the Hamiltonian, and S the role of the Lagrangian.

The front speed is bounded from above and below in terms of S.

Theorem 5.7 (Upper bound on front speed). *Let $b(y,t,\hat{\omega})$ satisfy assumptions A1–A6. Let $u(z,t,\hat{\omega})$ solve (5.52) with initial condition $u(z,0,\hat{\omega}) = u_0(z)$, where $u_0(z) \in [0,1]$ has compact support and is independent of $\hat{\omega}$. Then for any closed set $F \subset \{c \in \mathbb{R}^n \mid S(c) - f'(0) > 0\}$,*

$$\limsup_{t \to \infty} u(ct,t,\hat{\omega}) = 0$$

uniformly in $c \in F$, for almost every $\hat{\omega} \in \hat{\Omega}$.

The lower bound is given by the following theorem:

Theorem 5.8 (Lower bound on front speed). *Let $b(y,t,\hat{\omega})$ satisfy assumptions A1–A6. Let $u(z,t,\hat{\omega})$ solve (5.52) with initial condition $u(z,0,\hat{\omega}) = u_0(z)$, where $u_0(z) \in [0,1]$ has compact support and is independent of $\hat{\omega}$. Then for any compact set $K \subset \{c \in \mathbb{R}^n \mid S(c) - f'(0) < 0\}$,*

$$\liminf_{t \to \infty} u(ct,t,\hat{\omega}) = 1 \tag{5.59}$$

uniformly in $c \in K$, for almost every $\hat{\omega} \in \hat{\Omega}$.

Putting the upper and lower bounds together, we see that c^* in the unit direction $e \in \mathbb{R}^n$ satisfies the equation

$$S(c^*e) = f'(0), \tag{5.60}$$

implying, in terms of the "Hamiltonian" μ (following the remarks after Theorem 2.5), the *front speed variational formula*

$$c^*(e) = \inf_{\lambda \cdot e > 0} \frac{\mu(\lambda) + f'(0)}{\lambda \cdot e}. \tag{5.61}$$

Although the solution u depends on $\hat{\omega} \in \hat{\Omega}$, since B is a random variable over $\hat{\Omega}$, the function $S(c)$ and the speeds $c^*(e)$ are independent of $\hat{\omega}$. They are almost surely constant with respect to $\hat{Q}$, a consequence of the ergodicity assumption A3. Ergodicity follows from mixing or sufficient decay of correlations, so A5 would imply A3 if $\hat{\Gamma}$ were assumed to decay fast enough in large r. The assumption A2 (stationarity) is needed here so that c^* becomes constant as in stochastic homogenization in Chapter 4. Assumption A6 is a regularity condition, which also appears in analysis of Chapters 3 and 4. The Lipschitz assumption of b in space is needed to define unique solutions (characteristics) to the Itō equation (5.55). The periodicity of $b(y,t)$ in y is to provide compactness in the y dimensions. The results with minor modification extend to fronts in an infinite cylinder with the zero Neumann boundary condition on the sides of the cylinder, the setting considered [31].

In the infinite cylinder, the compactness property remains, and the process $Z^{z,t}(s)$ just needs to be reflected when it hits the boundary $\mathbb{R} \times \partial\Omega$. It is also possible to treat the non-Gaussian process $b(y,t)$ in the same framework as long as the process satisfies a few estimates in the proof [179].

Now let us relate the μ formula of Proposition 5.6 to a stochastic PDE. First, we note that $Z^{z,t} - z$ is the deviation of $Z^{z,t}(s)$ from the starting point. Without loss of generality, let $n = 2$. By stationarity, we may set $z = (x,y) = (0,0)$. The expectation term is

$$E\left[e^{-\lambda \cdot Z^{0,t}}\right] = E\left[e^{-\lambda_1 W_{1,t} - \lambda_1 \sigma \int_0^t \xi(W_{1,s}, t-s)\, ds}\right] \cdot E\left[e^{-\lambda_2 W_{2,t}}\right],$$

where we have used the independence of W_1 and W_2. The second factor equals $\exp\{\lambda_2^2 t/2\}$ by a direct calculation on the Gaussian random variable $W_{2,t}$ (mean zero, variance t). The first factor is the Feynman–Kac formula and equals $u(0,t)$, where $u = u(y,t)$ is the solution of

$$u_t = \frac{1}{2} u_{yy} - \lambda_1 b(y,t) u, \quad y \in \mathbb{R}^1,$$
$$u(y,0) = e^{-\lambda_2 y}. \tag{5.62}$$

The solutions of (5.62) are invariant in the sense of distributions if $(\lambda_1, \lambda_2) \to -(\lambda_1, \lambda_2)$, $x \to -x$, implying that $\lim_{t\to\infty} t^{-1} \log u(0,t) = \mu(\lambda)$ is an even function in λ. The function $v = \log u(y,t)$ satisfies the viscous Hamilton–Jacobi equation

$$v_t = \frac{1}{2} v_{yy} + \frac{1}{2} v_y^2 - \lambda_1 b(y,t), \tag{5.63}$$

with linear initial data $v(y,0) = -\lambda_2 y$. The limit $\lim_{t\to\infty} t^{-1} v$ agrees with the homogenized Hamiltonian. We see that μ is a convex function in λ_2, being the homogenization limit of a quadratic Hamiltonian with oscillating potential (for fixed λ_1). Evenness and convexity of μ in λ_2 imply that $\mu(\lambda_1,0) \leq \mu(\lambda_1,\lambda_2)$, and so the variational formula (5.61) reads, for $e = (1,0)$,

$$c^* = \inf_{\lambda_1 > 0} \frac{\mu(\lambda_1,0) + f'(0)}{\lambda_1}. \tag{5.64}$$

At $\lambda_2 = 0$, the problem (5.62) reduces to the form of the *parabolic Anderson model*, studied extensively in [48, 50, 62, 101]. The asymptotic growth rate of solutions

$$\lim_{t\to\infty} t^{-1} \log u(0,t) = \gamma, \tag{5.65}$$

if it exists and is nonrandom, is called the almost sure Lyapunov exponent. In the study of the parabolic Anderson model, b is a random potential in both y and t. Results are mostly in the regime of b being white noise in time or independent of time (stationary). In the stationary case $b = b(y,\omega)$, the limit (5.65) is closely related to the geometry of high peaks of the potential b and the corresponding peaks of solutions. At large time, the solution exhibits a spatially extremely irregular structure consisting of islands of high peaks located far from each other. This is reflected in the fact that first moment of solution $E[u(x,0)]$ grows much faster than individual realizations $u(0,t)$ as $t \to +\infty$. To leading order,

$$t^{-1} \log u(0,t) \sim \max_{|y| \le t} |\lambda_1| b(y, \omega), \tag{5.66}$$

as shown in [49, 101]. The viscous term enters the next-order asymptotics. Hence there is divergence of front speed c^* in KPP or viscous HJ due to unbounded running maxima in time-independent Gaussian shear flows, first observed in [247]. For spatial Gaussian processes (such as O-U) [50], the leading-order exponential growth rate of u has the asymptotic t-dependence

$$\sup_{|x| \le t} \xi(x) \sim \sqrt{2\Gamma(0) \log t}, \quad t \to \infty, \text{ a.s.,} \tag{5.67}$$

$\Gamma(0) = E[b^2]$. So

$$\lim_{t \to \infty} \frac{1}{t \sqrt{\log t}} \log u(0,t) = |\lambda_1| \sqrt{2\Gamma(0)}, \text{ a.s.,} \tag{5.68}$$

implying front-speed divergence in time. At large t, we have formally

$$\frac{\mu(\lambda_1, 0) + f'(0)}{\lambda_1} \sim \frac{f'(0)}{\lambda_1} + \frac{\lambda_1}{2} + \text{sign}(\lambda_1) \sqrt{2\Gamma(0) \log t},$$

or

$$c^* = \inf_{\lambda_1 > 0} \frac{\mu(\lambda_1, 0) + f'(0)}{\lambda_1} \sim c_0 + \sqrt{2\Gamma(0) \log t}. \tag{5.69}$$

The breakdown phenomenon of stochastic homogenization of inviscid quadratic HJ in Chapter 4 extends to viscous quadratic HJ. The KPP front speed c^* is related to stochastic homogenization of viscous HJ. See [135, 150] for the existence of the homogenization limit for convex and bounded spatially random Hamiltonians, and [136] for convex and bounded space–time random Hamiltonians.

In the time-dependent (white noise) case, the sign change of the potential promotes mixing and does not support growth of large peaks in solutions. The almost sure limit (5.65) is finite in both discrete and continuous models [48, 62]. In our case, we handle correlated noise in time for y in a bounded domain. The proof of the existence of μ uses the subadditive ergodic theorem on the logarithm of the infimum and supremum over z of the expectation term in (5.57), then a Harnack inequality to show that the two limits from inf and sup are the same. The subadditivity follows from the Markov property (independent increment) of the Wiener process.

Proof of the KPP front speed depends on large-deviation estimates of the random variable

$$\eta_z^t(\kappa t) = \frac{z - Z^{z,t}(\kappa t)}{\kappa t}, \tag{5.70}$$

which is the average velocity of a trajectory over the time interval $[0, \kappa t]$. The need for the parameter $\kappa \in (0, 1]$ results from treating the time-dependence of the field $b(y,t)$. Then the work is to show that almost surely, the random variables $\eta_z^t(\kappa t)$ satisfy a large-deviation principle (LDP) with convex and superlinear rate function $S(c)$ that is nonrandom and independent of z. The following properties are valid:

(i) For each $s \geq 0$, the set $\Phi(s) = \{c \in \mathbb{R} \mid S(c) \leq s\}$ is compact.
(ii) For any $\delta, h > 0$, there exists $t_0 > 0$ such that for all $t > t_0$

$$P\left(d(\eta_z^t(\kappa t), \Phi(s)) > \delta\right) \leq e^{-\kappa t(s-h)}.$$

(iii) For any $\delta, h > 0$, there exists $t_0 > 0$ such that for all $t > t_0$

$$P\left(\eta_z^t(\kappa t) \in U_\delta(c)\right) \geq e^{-\kappa t(S(c)+h)}. \tag{5.71}$$

The decoupling of two components of $Z^{z,t}(s)$ in the shear flow and the integral representation of Itō solutions (5.56) are used to prove the LDP [179].

The Feynman–Kac representation for u to be used with the large-deviation estimate is

$$u(z,t) = E\left[e^{\int_0^t \zeta(t-s,u(Z^{z,t}(s),t-s))ds}u_0(Z^{z,t}(t))\right], \tag{5.72}$$

or

$$u(z,t) = E\left[e^{\int_0^{t\wedge\tau} \zeta(t-s,u(Z^{z,t}(s),t-s))ds}u(Z^{z,t}(t\wedge\tau),t-(t\wedge\tau))\right], \tag{5.73}$$

where τ is a stopping time, the expectation is on the diffusion process $Z^{z,t}$, and $\zeta(u) = f(u)/u$. Roughly speaking, for large t, the expectation is supported on those paths $Z^{z,t}$ that travel a distance of $O(t)$ and pass through regions where u is small ($\zeta(u)$ is nearly maximized) but $u(Z^{z,t}(t-(t\wedge\tau)),t-(t\wedge\tau))$ is not too small. The lower bound is obtained by carefully choosing stopping times and estimating the probability of such paths.

The large-deviation estimate applied to the Feynman–Kac solution formula then implies that for any compact set $K \subset \{c \in \mathbb{R}^n \mid S(c) - f'(0) > 0\}$,

$$\liminf_{t\to\infty} \frac{1}{t} \log \inf_{c\in K} u(ct,t) \geq -\max_{c\in K}(S(c) - f'(0)), \tag{5.74}$$

holds almost surely. The estimate (5.74) is critical for proving the lower bound that

$$\liminf_{t\to\infty} u(ct,t) \geq 1 - h,$$
$$c\in K'$$

for any $h \in (0,1)$, where K' is any compact subset of $\{c \in \mathbb{R}^n : S(c) - f'(0) < 0\}$. The upper bound is more straightforward, where we replace $\zeta(u)$ by $f'(0)$ in the Feynman–Kac formula (5.72), so that the representation becomes explicit. Then we extract exponential asymptotics by large-deviation estimation (as shown in Chapter 2). The lower bound of u uses the Feynman–Kac formula with nontrivial stopping times (5.73) to allow conditioning on the Itō paths $Z^{z,t}(s)$ to pass through where u is small in a controllable manner.

5.2.2 Speed Bounds and Asymptotics

Variational formula (5.61) allows us to bound c^*. Throughout this section we assume that the shear $b(y,t)$ has the form

$$b(y,t) = \sum_{j=1}^{N} b_1^j(y)b_2^j(t), \tag{5.75}$$

where $b_1^j(y)$ are Lipschitz continuous and periodic in y, and $b_2^j(t)$ are stationary centered Gaussian fields such that assumptions A1–A6 are satisfied. We consider only front propagation in the direction $k = (1,0)$, which is aligned with the direction of the shear. The variational formula for the front speed reduces to the one-dimensional optimization problem

$$c^*(k) = \inf_{\lambda_1 > 0} \frac{\mu(\lambda_1, \mathbf{0}) + f'(0)}{\lambda_1},$$

where $\mu(\lambda_1, \mathbf{0})$ is determined by the limit

$$\mu(\lambda_1, \mathbf{0}) = \frac{\lambda_1^2}{2} + \lim_{t \to \infty} \frac{1}{t} \log \phi(y,t) \tag{5.76}$$

and

$$\phi_t = \frac{1}{2}\Delta_y \phi - \lambda_1 \, b(y,t) \, \phi, \qquad \phi|_{t=0} \equiv 1. \tag{5.77}$$

We consider the scaling $b(y,t) \mapsto \delta b(y,t)$ and the resulting enhancement of the corresponding speed $c^* = c^*(\delta)$. The analytical bounds below in particular establish the linear growth law for large δ, extending corresponding results in periodic media (Chapter 2).

Theorem 5.9 (Bounds on c^*). *For all $\delta \geq 0$, $c^*(\delta)$ satisfies the bounds*

(i) $c^*(\delta) \geq c^*(0)$.

(ii) $c^*(\delta) = c^*(0)$ *if* $b(y,t) = b(t)$.

(iii) $c^*(\delta) \leq c^*(0) + \delta \sum_{j=1}^{N} \|b_1^j\|_\infty E_Q[|b_2^j|]$.

(iv) $c^*(\delta) \leq c^*(0)\sqrt{1 + \delta^2 p_1}$, *where, $p_1 = \int_0^\infty \hat{\Gamma}(r)\,dr$ is the integral of temporal correlation of the shear in assumption A5.*

It follows from statement (iv) that when δ is small,

$$c^*(\delta) \leq c^*(0)\left(1 + \frac{\delta^2 p_1}{2}\right) + O(\delta^3),$$

the upper bound for the quadratic law.

Theorem 5.10 (Linear growth of c^*). *The nonrandom constant $\bar{C} \in [0,+\infty)$ defined by*

$$\liminf_{\delta \to \infty} \frac{c^*(\delta)}{\delta} = \bar{C} \tag{5.78}$$

is equal to zero if and only if $b(y,t) \equiv b(t)$.

Let us prove the inequality (iv) above, which uses the Gaussian property of the shear. Recall that the exponential term $e^{-\lambda \cdot (Z^{z,t}(t)-z)}$ in the μ formula (5.57) equals

$$\xi(t,W) \equiv e^{-\int_0^t \lambda_1 b(W_2(s)+z,t-s)\,ds - \lambda_1 W_1(t) - \lambda_2 \cdot W_2(t)}.$$

For any fixed continuous path W, $\xi(t,W)$ is lognormal with mean

$$E_Q[\xi(t,W)] = e^{|\lambda|^2 \hat{\sigma}^2/2} e^{-\lambda_1 W_1(t) - \lambda_2 \cdot W_2(t)}, \tag{5.79}$$

where

$$\hat{\sigma}^2 = \int_0^t \int_0^t \Gamma(W(s),W(r),s,r)\,ds\,dr \tag{5.80}$$

$$\leq \int_0^t \int_0^t \sup_{y_1,y_2} \Gamma(y_1,y_2,s,\tau)\,ds\,d\tau$$

$$\leq 2 \int_0^{\sqrt{2}t} \int_0^{t/\sqrt{2}} \hat{\Gamma}(r)\,dr\,dv$$

$$\leq 2\sqrt{2} p_1 t, \tag{5.81}$$

where we have made the change of variables $r = s - t$, $v = s + t$. Under the scaling $b \mapsto |\lambda|\delta b$, the constant p_1 defined in assumption A5 is replaced by $p_1 \mapsto |\lambda|^2 \delta^2 p_1$. Taking expectation $E_P[\cdot]$ of W in the expression of (5.79) shows that

$$\mu(\lambda) \leq |\lambda|^2/2 + \sqrt{2}|\lambda|^2 \delta^2 p_1,$$

and so

$$c^*(\delta) = \inf_{\lambda_1 > 0} \frac{\mu(\lambda_1, 0) + f'(0)}{\lambda_1} \leq \inf_{\lambda_1 > 0} \frac{\lambda_1}{2} + \frac{f'(0)}{\lambda_1} + \frac{\lambda_1^2 \delta^2 p_1}{2}$$

$$= 2\sqrt{(1 + \delta^2 p_1)f'(0)/2} = c^*(0)\sqrt{1 + \delta^2 p_1}. \tag{5.82}$$

The upper bound (iv) immediately yields a proof of speed-bending studied in the combustion literature [65]. Experiments with premixed flames have shown that increasing turbulence intensity does not lead to unlimited linear enhancement of the turbulent burning rate [203]. It has been proposed [65] that this "bending" of the turbulent burning velocity in high-intensity flows can be explained by a rapid temporal decorrelation of the flow. The simulation of [65] used a viscous G-equation. For the KPP model, the following upper bound confirms that rapid temporal decorrelation also leads to sublinear enhancement of the KPP front speed. The agreement of the

G-equation and the KPP model predictions identifies the time decorrelation of the flow field as one mechanism of speed-bending among others [65].

Notice that the derivation uses no information about the spatial structure of the flow other than the maximum value $\|b_1^j\|_\infty$. As a result, it is likely that the actual speed may grow more slowly than $\delta^{1/2}$, or that c^* eventually decreases with δ, as suggested by numerical experiments of [65].

Corollary 5.11 (Bending of Front Speed). *For $\delta > 0$, let $\{b_2^j(t)\}_{j=1}^N$ be a family of stationary Gaussian fields on $[0, \infty)$ satisfying $E_Q[b_2^j(s)b_2^k(t)] \leq C_1 e^{-\alpha_{j,k}|t-s|}$, where $\alpha_{j,k} > 0$ and $C_1 > 0$. Then for the scaled flow $b^\delta(y,t) = \sum_{j=1}^N \delta b_1^j(y)b_2^j(\delta t)$,*

$$\limsup_{\delta \to \infty} \frac{c^*(\delta)}{\sqrt{\delta}} < +\infty. \tag{5.83}$$

Proof. For the flow $\sum_{j=1}^N b_1^j(y)b_1^j(\delta t)$, we have

$$\begin{aligned}
\hat{\Gamma}(r) &= \sup_{y_1, y_2} \Gamma(y_1, y_2, 0, r) \\
&\leq \sum_{j,k} \|b_1^j\|_\infty \|b_1^k\|_\infty E_Q[b_2^j(0)b_2^k(\delta r)] \\
&\leq \sum_{j,k} \|b_1^j\|_\infty \|b_1^k\|_\infty C_1 e^{-\alpha_{j,k}\delta|r|},
\end{aligned} \tag{5.84}$$

and so

$$p_1 = \int_0^\infty \hat{\Gamma}(r)\, dr \leq \delta^{-1} C_1 \sum_{j,k} \frac{\|b_1^j\|_\infty \|b_1^k\|_\infty}{\alpha_{j,k}}.$$

The front-bending result (5.83) now follows from part (iv) of Theorem 5.9. $\qquad\square$

The quadratic and linear growth laws can be formally derived for KPP front speeds in a stationary space–time Gaussian white-in-time shear field $b(y,t,\omega)$ over $\mathbb{R}^n$ based on the asymptotic properties of the Lyapunov exponent [247]. Let the co-variance function of b be $\Gamma(t,y) = \delta(t)\Gamma_0(|y|)$. Asymptotic results for the principal Lyapunov exponent of the parabolic Anderson problem can be stated as follows [48, 62]. Let v be the solution of

$$v_t = \kappa \Delta v - b(y,t)v, \quad x \in \mathbb{R}^n, \ \kappa > 0. \tag{5.85}$$

Then the almost sure Lyapunov exponent $\gamma(\kappa)$ exists as a nonrandom number and obeys ($n = 1, 2, \ c_1 > 0$)

$$\gamma(\kappa) \sim c_1 \kappa^p, \quad p \in (0, 1/2), \ \kappa \ll 1, \tag{5.86}$$

$$\gamma(\kappa) \sim \frac{\Gamma_0(0)}{2}, \quad \kappa \gg 1. \tag{5.87}$$

Moreover, $\gamma(\kappa)$ is a monotone increasing continuous function in κ [48]. The results can be adapted to equation (5.77) with $b \mapsto \delta b$. First $\frac{1}{\sqrt{a}} b(y,t/a) = b(y,t)$ in law, for any $a > 0$. Let $t = t'/a$, $a = (\delta \lambda_1)^2$. Then equation (5.62) becomes

$$u_{t'} = \frac{1}{2\delta^2 \lambda_1^2} \Delta u - b(y, t')u. \tag{5.88}$$

It follows that $\lim_{t \to \infty} \log u(0, t)/t = \gamma^* = \gamma^*(\delta \lambda_1)$ such that

$$\gamma^*(\delta \lambda_1) \sim \frac{\Gamma(0)(\delta \lambda_1)^2}{2}, \quad \delta \lambda_1 \ll 1, \tag{5.89}$$

$$\gamma^*(\delta \lambda_1) \sim b_0 h \left(\frac{1}{2\delta^2 \lambda_1^2} \right) (\delta \lambda_1)^2, \quad \delta \lambda_1 \gg 1, \tag{5.90}$$

where $h = h(x) = x^p$, $p \in (0, 1/2)$.

Now we minimize $\mu(\lambda_1, 0)/\lambda_1$. For $\delta \ll 1$, we have

$$\frac{\mu(\lambda_1, 0)}{\lambda_1} \sim \frac{f'(0)}{\lambda_1} + \frac{\lambda_1}{2} + \frac{\Gamma(0)\delta \lambda_1}{2} + \text{higher-order terms},$$

giving minimum value

$$c^* = c^*(0) \left(1 + \frac{1}{2} \Gamma(0)\delta^2 \right) + \text{higher-order terms}. \tag{5.91}$$

This is the stochastic analogue of quadratic speed enhancement. It is interesting that in this case, the integral of (5.80) equals $\Gamma(0)t$, and (5.91) follows also from the μ formula of Proposition 5.6. The linear law in large δ follows from (5.90) and the speed variational formula [247].

5.3 KPP Fronts in Spatially Random Compressible Flows

In the previous two sections, the randomness in the flow appears in time or in a direction orthogonal to that of front propagation. The front speeds are enhanced due to shear structure of the flows. In this section, we study the case in which randomness is in the direction of front propagation. We consider solutions to the KPP reaction–advection–diffusion equation

$$u_t = \frac{1}{2} u_{xx} + b(x)u_x + f(u), \quad t > 0, \ x \in \mathbb{R}. \tag{5.92}$$

The initial datum $u_0(x) \in [0, 1]$ is compactly supported. The advection $b(x)$ under our assumptions below will be compressible.

The pioneering work on KPP front speeds in random media was based on the large-deviation method [94, 100] in the late 1970s for the KPP equation

$$u_t = u_{xx} + f(x, u), \quad x \in \mathbb{R}, \tag{5.93}$$

where $f(x,u)$ is a KPP nonlinearity with random stationary ergodic dependence in x, and the initial data are nonnegative and compactly supported. The variational formula of front speeds was proved [100]; see also [96, Section 7.7] for an example of two-state Markov dependence of f in x and the representation of the variational formula in special functions. The next result along this line was [144], analyzing a d-dimensional ($d \geq 2$) lattice KPP equation of the form

$$u_t = \tilde{\Delta} u + \xi(x)u(1-u), \quad t > 0, \ x \in Z^d, \tag{5.94}$$

with initial condition $u(0,x) = 1$ if $x = 0$, $u(0,x) = 0$ otherwise. Here

$$\tilde{\Delta} f(x) = \frac{1}{2d} \sum_{\substack{e \in Z^d \\ |e|=1}} f(x+e) - f(x),$$

the discrete Laplacian. The random variables $\xi(x)$ are independent and identically distributed, bounded, and nonnegative. A similar variational front speed formula holds.

The results in this section are concerned with front speeds in unbounded random advection by a combination of the large-deviation approach and the techniques from analysis of random walks in random environments. Moreover, we shall show nearly optimal asymptotic bounds on the front speeds for strong advection.

For the random drift $b(x,\hat{\omega}) : \mathbb{R} \times \hat{\Omega} \to \mathbb{R}$, we make the following assumptions:

(1) (stationarity) b is a stationary random process on $\mathbb{R}$ defined over the probability space $(\hat{\Omega}, \hat{\mathscr{F}}, Q)$ with zero mean, $E_Q[b] = 0$.
(2) (ergodicity and regularity) $b(\cdot, \hat{\omega})$ is almost surely locally Lipschitz continuous and translation with respect to x generates an ergodic transformation of the space $\hat{\Omega}$.
(3) (moment condition) the process $b(x, \hat{\omega})$ satisfies

$$E_Q \left[\sup_{x \in [-2,2]} |b(x, \hat{\omega})| \right] < \infty. \tag{5.95}$$

However, *we do not assume that the process b is globally bounded or globally Lipschitz continuous.*
(4) For some $\alpha_1, \alpha_2 \in \mathbb{R}$,

$$\limsup_{z \to \infty} Q \left(\int_0^z b(s, \hat{\omega}) \, ds \geq \alpha_1 \right) < 1 \tag{5.96}$$

and

$$\limsup_{z \to \infty} Q \left(\int_{-z}^0 b(s, \hat{\omega}) \, ds \leq \alpha_2 \right) < 1. \tag{5.97}$$

Assumption (4) is not restrictive. For example, if $b(x, \hat{\omega})$ is square integrable and sufficiently mixing with respect to shifts in x, then b satisfies an invariance principle

[36]

$$\frac{1}{\sigma\sqrt{z}} \int_0^z b(x,\hat{\omega})\,ds \to N(0,1), \quad Q\text{-a.s.}, \tag{5.98}$$

which implies (5.96). In particular, all the above assumptions on b hold for a mean-zero locally Lipschitz continuous Gaussian process with sufficient decay of correlation functions, while (5.95) follows from the Borel inequality [4]. The Borel inequality for Gaussian fields on $\mathbb{R}^n$ states that if $\|f\| = \sup_{y\in M} b(y)$ is almost surely finite, M a compact subset of $\mathbb{R}^n$, then $E_Q[\|f\|] < \infty$; moreover, for any $u > 0$,

$$Q\left(|\,\|f\| - E[\|f\|]\,| > u\right) \le e^{-u^2/2\sigma_M^2}, \tag{5.99}$$

where $\sigma_M^2 = \sup_{y\in M} E_Q[b^2(y)]$.

We now present some of the main results of recent work [181] on front propagation and speeds bounds.

Theorem 5.12. *Suppose that assumptions (1)–(4) hold. Then there are deterministic constants $c_-^* < 0$ and $c_+^* > 0$ such that for any closed set $F \subset (-\infty, c_-^*) \cup (c_+^*, +\infty)$,*

$$\lim_{t\to\infty} \sup_{c\in F} u(ct, t, \hat{\omega}) = 0$$

for almost every $\hat{\omega} \in \hat{\Omega}$. Also, for any compact set $K \subset (c_-^*, c_+^*)$,

$$\lim_{t\to\infty} \inf_{c\in K} u(ct, t, \hat{\omega}) = 1$$

for almost every $\hat{\omega} \in \hat{\Omega}$.

The next result describes the effect of scaling the drift $b \mapsto \delta b$, where $\delta \in [0, \infty)$ is a scaling parameter. The front speed decreases to zero as the flow amplitude increases:

Theorem 5.13. *The front speed $c_+^*(\delta)$ satisfies the lower bound*

$$c_+^*(\delta) \ge \frac{1}{C} \min\left(1, \frac{f'(0)}{1+\delta M}\right), \tag{5.100}$$

where C is a universal constant and $M = E_Q\left[\sup_{x\in[-2,2]} |b(x,\hat{\omega})|\right]$. Moreover, for any $p \in (0,1)$, there is a random constant $C' = C'(p,\hat{\omega})$ such that

$$c_+^*(\delta) \le C'\delta^{-p} \tag{5.101}$$

for all $\delta > 0$. Therefore $\limsup_{\delta\to\infty} c_+^*(\delta) = 0$ holds with probability one. Similar statements hold for $|c_-^*(\delta)|$.

Our analysis of $u(x,t)$ involves large deviations estimates for the associated diffusion process $X^x(t)$ in the random environment. From assumption (5.95) and the assumption of stationarity and ergodicity, it follows that almost surely with respect to Q there is a constant $k = k(\hat{\omega})$ such that $|b(x,\hat{\omega})| \le k(1+|x|)$ for all $x \in \mathbb{R}$. Therefore, for each $\hat{\omega} \in \hat{\Omega}$ fixed, we can define $X^x(t)$ to be the strong solution to the Itô equation:

$$X^x(t) = x + \int_0^t b(X^x(s))\,ds + W(t), \qquad (5.102)$$

where $W(t) = W(t,\omega)$ is a one-dimensional Wiener process (Brownian motion) defined on $(\Omega, \mathscr{F}, P)$ with $W(0) = 0$, P-a.s. The large-deviation method for KPP fronts was developed for one-dimensional spatial random media in the late 1970s. Chapter VII of [96] analyzed equation (5.92) under the assumption that b is uniformly bounded or that randomness appears in the nonnegative reaction f. Solutions to (5.92) in multiple dimensions with uniformly bounded random coefficients were also studied recently [150] as an application of stochastic homogenization of viscous HJ. See also [135] for a different method of HJ homogenization. Our results treated unbounded advection and provided concrete tight bounds on front speeds in the limit of large advection. Let us return to the HJ perspective of KPP to explain why unboundedness of b does not cause front speed to diverge. The unbounded advection is a gradient flow, and the corresponding Hamiltonian is $H(p,x) = p^2/2 + b(x)p$. As we discussed in Example 4.10, the HJ asymptotic front speed is finite in spite of the unboundedness of b, just as here.

The bounds (5.100)–(5.101) tell us about the rate of "front trapping" by random media in the large-advection limit. A few remarks are in order about front speeds and diffusion in random media. When $b \equiv 0$, the solutions to the initial value problem (5.92) develop fronts propagating with speed equal to $c^* = 2\sqrt{\kappa f'(0)}$, where κ is the diffusion constant ($\kappa = 1/2$ in (5.92)). This suggests that for nonzero b (stationary and ergodic) one might estimate the front speed by $c^* \approx c^{\bar{\kappa}} = 2\sqrt{\bar{\kappa}f'(0)}$, where $\bar{\kappa} = \lim_{t\to\infty} E[|X^x(t)|^2]/t$ is the Lagrangian way of writing the homogenized (effective) diffusivity of the random medium. For periodic incompressible two-dimensional velocity fields, the ratio $c^*/c^{\bar{\kappa}}$ is bounded away from zero and infinity by constants independent of the flow [208]. In the random setting here, however, even when front speeds and $\bar{\kappa}$ are both finite, such equivalence does not hold. The two quantities $c^{\bar{\kappa}}$ and c^* may scale quite differently with respect to δ. When b is uniformly bounded, the process $X^x(t)$ is diffusive [190, 214] with effective diffusivity

$$\bar{\kappa} = \frac{1}{E_Q[e^{-b}]E_Q[e^b]} > 0, \qquad (5.103)$$

almost surely with respect to Q. Suppose that the distribution of b is sign-symmetric (i.e., $b \overset{L}{=} -b$). Then $E_Q[e^{-b}] = E_Q[e^b]$, so that the effective diffusivity is

$$\bar{\kappa} = \frac{1}{(E_Q[e^b])^2}.$$

In this case, the effective diffusivity (and $c^{\bar{\kappa}}$) will decrease exponentially fast as the scaling parameter δ is increased. However, the lower bound in Theorem 5.13 shows that the corresponding front speed can decrease no faster than $O(\delta^{-1})$ as δ

increases. The reason for this difference is that the front speed is determined by large deviations of the diffusion process $X^x(t)$, which may not be accurately predicted by the asymptotic behavior of the variance of the process.

Because $X^x(t)$ does not have an explicit formula in terms of b (compare equation (5.102) with equation (5.56)), large-deviation estimates for $X^x(t)$ are derived for the hitting time T_r^s, the first time the process hits the point $x = r$ (from the right) starting from $x = s \geq r$. Then the hitting time estimates are translated into those for the average velocity of trajectory $X^x(t)$. For the right-moving front (left-moving fronts are similar), we have the following result:

Proposition 5.14. *Suppose $b(x, \hat{\omega})$ satisfies assumptions (1)–(4). Almost surely with respect to Q, the following estimates hold. Let $v \in \mathbb{R}$, $\kappa \in (0,1]$. For any closed set $G \subset [(a_0)^{-1}, \infty)$, $a_0 > 0$ and deterministic,*

$$\limsup_{t \to \infty} \frac{1}{\kappa t} \log P\left(\frac{vt - X^{vt}(\kappa t)}{\kappa t} \in G \right) \leq -\inf_{c \in G} cI^+\left(\frac{1}{c}\right), \qquad (5.104)$$

and for any open set $F \subset [(a_0)^{-1}, \infty)$,

$$\liminf_{t \to \infty} \frac{1}{\kappa t} \log P\left(\frac{vt - X^{vt}(\kappa t)}{\kappa t} \in F \right) \geq -\inf_{c \in F} cI^+\left(\frac{1}{c}\right). \qquad (5.105)$$

Here the function $I^+(a)$ is deterministic and satisfies

(i) *$I^+(a) > 0$ for $a \in (0, a_0)$, $a_0 \in (0, \infty]$.*

(ii) *$I^+(a)$ is convex and decreasing in a for $a \in (0, a_0)$.*

(iii) *$\lim_{a \to 0^+} I^+(a) = +\infty$, and $\lim_{a \to (a_0)^-} I^+(a) = 0$.*

(iv) *If $a_0 < \infty$, then $I^+(a_0) = 0$, and $I^+(a) \geq 0$ for $a \in (a_0, \infty)$.*

Define nonrandom constants $c_+^* > 0$ by the equation

$$(c_+^*)I^+(1/c_+^*) = f'(0). \qquad (5.106)$$

Then by the properties of $I^+(a)$, c_+^* exists uniquely, $1/c_+^* \in (0, a_0)$, and $cI^+(1/c) > f'(0)$ for all $c > c_+^*$. The function $cI^+(1/c)$ is the analogue of $S(c)$, the Lagrangian (action) in the analysis of shear flows. The large-deviation upper bound above (5.104) gives, via the Feynman–Kac formula,

$$\lim_{t \to \infty} \frac{1}{t} \sup_{c \in F} \log u(ct, t) \leq f'(0) - cI^+\left(\frac{1}{c}\right) < 0,$$

for $c \in F \subset (c_+^*, \infty)$. The large-deviation lower bound above (5.105) leads to the lower bound of solution u:

$$\liminf_{t \to \infty} \frac{1}{t} \log \inf_{c \in K} u(ct, t) \geq -\max_{c \in K}\left(cI^+\left(\frac{1}{c}\right) - f'(0) \right), \qquad (5.107)$$

almost surely in Q, for any compact set $K \subset (c_+^*, \infty)$.

The propagation speed c_+^* is then justified with these bounds of solutions as before. The speed bounds come from analyzing further the I^+ function and the hitting times of the processes $X^x(t)$ using properties of b and the Wiener process [181].

A related class of random KPP equations arising from the limit of certain interacting particle system are of the type

$$u_t = u_{xx} + a(u) + \varepsilon b(u)\dot{W}, \qquad (5.108)$$

where $a(u)$ is a KPP nonlinearity, typically equal to $u(1-u)$; $b(u)$ is a Lipschitz continuous function of u; and $\dot{W} = \dot{W}(x,t)$ is a space–time white noise. Random traveling fronts [166] are investigated in a special form of equation (5.108):

$$u_t = u_{xx} + u - u^2 + \varepsilon\sqrt{u(1-u)}\dot{W}, \qquad (5.109)$$

with the initial datum $u_0 = I_{(-\infty,a)}$.

Equation (5.109) is closely related to the so-called historical process; see [64] for details. Simply put, the historical process is a measure on the sets of paths over a time period $[0,t_0]$, which represent the past history up to time t_0 of a cloud of infinitesimally small particles whose density at time t is $u(x,t)$. The particles move according to independent Brownian motions, and they give birth and die. There is an excess of births over deaths, which has a size of $1-u$. An interesting property of this process is that a small collection of particles dies out quickly for large values of x for the given initial density u_0, which implies that the solution u has compact support on the positive x-axis. By symmetry of the solutions, u is also equal to 1 beyond a compact interval. The front location is defined as

$$b(t) = \sup\{x \in \mathbb{R} : u(x,t) > 0\}. \qquad (5.110)$$

The main results proved in [166] are given in the following theorem:

Theorem 5.15. *Consider a solution u of (5.109) with initial data $I_{(-\infty,a)}$, $a > 0$. With probability one, $0 \le u \le 1$ for all (x,t). For ε small enough, the solution u behaves like a moving front with the following properties:*

1. *(Front speed and shape) With probability one, $\lim_{t\to\infty} b(t)/t = c^*$ exists and $c^* \in (0,+\infty)$, and this limit depends on ε. The law of the front profile $v(x,t) = u(b(t)+x,t)$ tends toward a stationary limit as $t \to \infty$.*
2. *(Front width) Let $I(t) = [a(t),b(t)]$ be the smallest closed interval such that $u = 1$ for $x < a(t)$, and $u = 0$ if $x > b(t)$. Then with probability one, $I(t)$ is a compact interval for all $t \ge 0$.*

Theorem 5.15 seems to be the first on KPP random fronts providing information about front shape and width in addition to the front speed. The almost sure finite front width property is reminiscent of the viscous Burgers front under white noise perturbation.

Recently, the asymptotic behavior of c^* in ε has been obtained [58].

Theorem 5.16. *Under the conditions of Theorem 5.15, the front speed c^* satisfies the inequalities*

$$\liminf_{\varepsilon \to +\infty} \varepsilon^2 c^*(\varepsilon) \geq 2 \qquad (5.111)$$

and

$$2 \geq c^*(\varepsilon) \geq 2 - K \frac{\log\log(1/\varepsilon)}{[\log \varepsilon]^2} \qquad (5.112)$$

for $\varepsilon \in (0, 1/10)$ and some constant $K > 0$.

Inequality (5.111) says that $c^*(\varepsilon) > 0$ no matter how large ε is; hence there is no trapping of fronts. This is similar to KPP fronts in drift, or inequality (5.100). Numerical simulations [68] suggest that (5.111) is an equality, or $\lim_{\varepsilon \to +\infty} \varepsilon^2 c^*(\varepsilon) = 2$. This means that the space–time white noise slows down the front when ε is large, similar to KPP front speeds in large drift (5.101) except that the slowdown in (5.109) is faster $(O(\varepsilon^{-2}))$ as ε increases.

5.4 KPP Fronts in Space–Time Random Incompressible Flows

Reaction–diffusion front propagation in incompressible space–time random flows is a fundamental subject in premixed turbulent combustion [56, 251, 203, 153, 246]. The challenging mathematical problem is to establish the propagation velocity of the front (large-time asymptotic spreading rate) using the governing partial differential equations. Another mathematical problem is to characterize the propagation velocity, the turbulent flame speed [203], in terms of flow statistics. Due to the notorious closure problem in turbulence, the turbulent front speed has been approximated by ad hoc and formal procedures in the combustion literature, such as various closures and renormalization group methods [194, 251, 54].

However, these methods are difficult to justify mathematically. In the KPP model, a rigorous mathematical theory of front spreading and a variational formula of speed c^* can be achieved [182]. In other words, the large-time KPP front speed in space–time random incompressible flows is exactly solvable. We shall outline the main ingredients of the mathematical theory. Though turbulent combustion problems may be posed in terms of different models in the literature, the fundamental mathematical issues are the same. Solving KPP provides the first step toward pursuing other models.

5.4.1 Eulerian Method of Front Speeds in Random Flows

The KPP reaction–diffusion–advection equation is

$$\partial_t u = \Delta u + V(x, t, \hat{\omega}) \cdot \nabla u + f(u) \overset{\Delta}{=} \mathscr{L} u + f(u), \qquad (5.113)$$

with smooth, compactly supported, nonnegative initial data $u(x,0,\hat{\omega}) = u_0(x)$, $0 \leq u_0 \leq 1$. The vector field $V(x,t,\hat{\omega})$ is defined over a probability space $(\hat{\Omega}, \hat{\mathscr{F}}, \hat{P})$. We make the following assumptions:

(1) (Stationarity and ergodicity) The field V is stationary with respect to shifts in x and t: there is a group of measure-preserving transformations $\tau_{(x,t)} : \hat{\Omega} \to \hat{\Omega}$ such that $V(x+h, t+r, \hat{\omega}) = V(x,t,\tau_{(h,r)}\hat{\omega})$, and τ acts ergodically on $\hat{\Omega}$.

(2) (Regularity) V is locally Hölder continuous, almost surely, in the sense that for each $T > 0$ there is $\alpha = \alpha(\hat{\omega}, T) \in (0,1)$ such that

$$\|V(\cdot,\cdot,\hat{\omega})\|_{C^\alpha(\mathbb{R}^d \times [0,T])} < \infty \qquad (5.114)$$

holds for almost every $\hat{\omega} \in \hat{\Omega}$.

(3) (Incompressibility) The field V is divergence-free, $\nabla \cdot V = 0$, in the sense of distributions, almost surely with respect to $\hat{P}$.

(4) (Moment condition) The field V satisfies the bound

$$\bar{V}_2 \overset{\Delta}{=} E_{\hat{P}} \left[\sup_{(t,x) \in [0,1] \times \mathbb{R}^d} |V(x,t)|^2 \right] < \infty. \qquad (5.115)$$

Condition (4) means that $V(x,t,\hat{\omega})$ is uniformly bounded in x for each fixed t and $\hat{\omega}$. However, we do not require that $V(x,t,\cdot) \in L^\infty(\hat{\Omega})$, so that V may become unbounded as $t \to \infty$. The Hölder regularity condition (2) is satisfied by turbulent flows [153, 246] and is a physical assumption for turbulent combustion problems [203, 194, 246].

For almost every $\hat{\omega}$, there exists a unique classical solution satisfying (5.52). Our main result [182] is the following theorem regarding the almost sure asymptotic behavior of the solution $u(x,t,\hat{\omega})$ as $t \to \infty$:

Theorem 5.17. *There are a convex open set $G \subset \mathbb{R}^d$ and a set of full measure $\hat{\Omega}_0 \subset \hat{\Omega}$, $\hat{P}(\hat{\Omega}_0) = 1$, such that the following limits hold for all $\hat{\omega} \in \hat{\Omega}_0$:*

$$\lim_{t \to \infty} \sup_{c \in F} u(ct,t) = 0 \qquad (5.116)$$

for any closed set $F \subset \mathbb{R}^d \setminus \bar{G}$ and

$$\lim_{t \to \infty} \inf_{c \in K} u(ct,t) = 1 \qquad (5.117)$$

for any compact set $K \subset G$.

Thus, the deterministic set $\{ct \in \mathbb{R}^d \mid c \in \partial G\}$ represents the spreading interface in an asymptotic sense, made precise by (5.116) and (5.117). The set G may be characterized in the following way. Let $\phi(x,t,\hat{\omega}) \geq 0$ solve the advection–diffusion equation $\partial_t \phi = \mathscr{L}\phi$ with initial condition $\phi(x,0,\hat{\omega}) = \phi_0(x) \geq 0$, where $\phi_0(x)$ is smooth, deterministic, and compactly supported.

Theorem 5.18. *The limit*

$$\mu(\lambda) = \lim_{t \to \infty} \frac{1}{t} \log \int_{\mathbb{R}^d} e^{\lambda \cdot x} \phi(x,t,\hat{\omega}) \, dx = \lim_{t \to \infty} \frac{1}{t} \log E_{\hat{P}}[e^{\lambda \cdot x} \phi(x,t,\hat{\omega})] \qquad (5.118)$$

exists almost surely with respect to $\hat{P}$. Moreover, $\mu(\lambda)$ is a finite convex function of $\lambda \in \mathbb{R}^d$, and is superlinear at large λ.

The function μ is the "Hamiltonian," and now the characterization of G is given by the "Lagrangian":

Theorem 5.19. *The set G described in Theorem 5.17 is given by*

$$G = \{c \in \mathbb{R}^d \mid S(c) \le f'(0)\}, \qquad (5.119)$$

where $S(c) = \sup_{\lambda \in \mathbb{R}^d} (\lambda \cdot c - \mu(\lambda))$ and $\mu(\lambda)$ is defined as in Theorem 5.18. It follows that the asymptotic front speed c^* in the direction $e \in \mathbb{R}^d$ is given by the variational formula

$$c^*(e) = \inf_{\lambda \cdot e > 0} \frac{\mu(\lambda) + f'(0)}{\lambda \cdot e}. \qquad (5.120)$$

For the KPP model, Theorems 5.17 and 5.19 address two open problems in turbulent combustion [203]: the existence of a well-defined turbulent front speed and the precise analytical characterization of the speed. In Theorem 5.18, one may normalize ϕ so that ϕ is the density for a probability measure on $\mathbb{R}^d$, for each fixed $\hat{\omega}$, and the theorem characterizes the asymptotic behavior of the tails of the distribution (large deviations from the mean behavior) almost surely with respect to the measure $\hat{P}$ on the velocity field. The function S in Theorem 5.19 is the rate function that governs these large deviations. Because S is related to the action functional, it may be viewed as a Lagrangian.

The quantity $\mu(\lambda)$ has another interpretation in terms of exponential growth of PDE solutions, or the almost sure (principal) Lyapunov exponent. It can be formulated as either an initial value problem of the related PDE as we have done in Section 5.2 or a terminal value problem below. Consider the function $\varphi^*(x, \tau; t, \hat{\omega})$, which solves the terminal value problem ($\tau \in (0, t)$):

$$\partial_\tau \varphi^* + \Delta \varphi^* - (V(x, \tau) - 2\lambda) \cdot \nabla \varphi^* + (|\lambda|^2 - \lambda \cdot V(x, \tau)) \, \varphi^* = 0, \qquad (5.121)$$

with linear terminal data $\varphi^*(x, t; t, \hat{\omega}) \equiv 1$, $x \in \mathbb{R}^d$. Then $\varphi^*(x, 0; t, \hat{\omega})$ grows exponentially in t with a rate equal to $\mu(\lambda)$:

Theorem 5.20. *If $\varphi^*(x, \tau; t, \hat{\omega})$ solves (5.121) with terminal data $\varphi^*(x, t, \hat{\omega}) \equiv 1$, then for any $r > 0$,*

$$\lim_{t \to \infty} \sup_{|x| \le rt} \left| \frac{1}{t} \log \varphi^*(x, 0; t, \hat{\omega}) - \mu(\lambda) \right| = 0 \qquad (5.122)$$

holds almost surely with respect to the measure $\hat{P}$.

The function $\mu(\lambda)$ is related to the effective Hamiltonian that arises from stochastic homogenization of "viscous" Hamilton–Jacobi equations [135, 136, 150]. The convex function $\mu(\lambda)$ in (5.118) is equal to an effective Hamiltonian $\bar{H}(\lambda)$. To see this, define the function $\eta^*(x, \tau; t, \hat{\omega}) = e^{\lambda \cdot y} \varphi^*(x, \tau; t, \hat{\omega})$, which satisfies $\partial_\tau \eta + \mathscr{L}^* \eta^* = 0$ for $\tau < t$ and terminal data $\eta^*(x, t; t, \hat{\omega}) = e^{\lambda \cdot y}$. Here $\mathscr{L}^* \eta^* = \Delta_x \eta^* - \nabla \cdot (V \eta^*)$ denotes the adjoint operator. For $\varepsilon > 0$ and $T > 0$, define

$$\zeta^\varepsilon(x, \tau; T, \hat{\omega}) = \varepsilon \log \eta^*(\varepsilon^{-1} x, \varepsilon^{-1} \tau; \varepsilon^{-1} T, \hat{\omega}).$$

Then ζ^ε solves the Hamilton–Jacobi equation

$$\partial_\tau \zeta^\varepsilon + \varepsilon \Delta \zeta^\varepsilon + |\nabla \zeta^\varepsilon|^2 - V \left(\frac{x}{\varepsilon}, \frac{\tau}{\varepsilon}, \hat{\omega} \right) \cdot \nabla \zeta^\varepsilon = 0, \quad \tau \in [0, T), \tag{5.123}$$

with terminal data $\zeta^\varepsilon(x, T; T, \hat{\omega}) = \lambda \cdot x$. For a velocity field $V(x, \tau, \hat{\omega})$ that is uniformly bounded in τ (i.e., $V \in L^\infty(\hat{\Omega}; L^\infty(\mathbb{R}^d)))$, the result of [136] implies that as $\varepsilon \to 0$, the function ζ^ε converges locally uniformly to a function $\zeta^0(x, \tau; t)$ that solves an effective Hamilton–Jacobi equation $\partial_\tau \zeta^0(z, \tau; t) + \bar{H}(\nabla \zeta^0) = 0$ with the same terminal data. The effective Hamiltonian $\bar{H}(\lambda)$ is a deterministic function. In particular, by choosing $T = 1$, we see that

$$\bar{H}(\lambda) = \lim_{\varepsilon \to 0} \zeta^\varepsilon(0, 0; 1, \hat{\omega}) = \lim_{\varepsilon \to 0} \varepsilon \log \eta^*(0, 0; \varepsilon^{-1}, \hat{\omega}) = \mu(\lambda)$$

holds almost surely with respect to $\hat{P}$.

Theorem 5.20 extends this connection to the case of velocity fields $V(x, t)$ that are not uniformly bounded in t, a case not covered by [135, 136, 150]. As discussed in Section 5.2, the time randomness helps mixing and homogenization of HJ, so one would anticipate that our results also hold if spatial boundedness of V is removed. We shall show a formal calculation of c^* for a space–time Gaussian and white-in-time velocity field V later.

Because of the rather general form of V in several space dimensions, neither the explicit representation nor a useful hitting time formulation is available for analyzing the Itō paths of the associated diffusion process (the stochastic characteristics). This makes proving the large-deviation principle (LDP) based on Itō solutions difficult, and is perhaps the reason why progress has been slow since the one-dimensional KKP front result appeared in the 1970s [94, 100, 96]. A method of proving LDP based on analyzing Itō solutions is Lagrangian. A way to handle a more general form of advection V is the new Eulerian approach that we developed in [182].

The first step is to use the Krylov–Safonov–Harnack inequality [137] to establish continuity estimates of the solution. Though the constants appearing in the Krylov–Safonov–Harnack inequality may be arbitrarily bad, they are well behaved "on average." We use this observation and the subadditive ergodic theorem to establish almost sure behavior of the tails of the linearized KPP equation.

The tails of these solutions in the large-time limit contain the information on the "Lagrangian" function S (the large-deviation rate function). To apply this property to the solution of the nonlinear equation and recover front speed c^*, we construct subsolutions (supersolutions) and use the comparison principle to bound solutions instead of estimating with Itō solutions and the Feynman–Kac formula.

Since the proof relies only on the Krylov–Safonov–Harnack inequality and the comparison principle, it applies readily to a large class of operators $\mathscr{L}$. In fact, all of the proofs may be modified slightly to treat the case that the diffusion is also a random process. For example, a variant of Theorems 5.17–5.20 holds in the case that u is governed by an equation of the form

$$\partial_t u = \nabla \cdot (A(x,t,\hat{\omega})\nabla u) + V(x,t,\hat{\omega}) \cdot \nabla u + f(u), \qquad (5.124)$$

where $A(x,t,\hat{\omega}) = A_{ij}(x,t,\hat{\omega})$ is a random positive definite matrix function and uniformly $C^{1,\alpha}$.

Let us walk through the main estimates. We begin with the Krylov–Safonov–Harnack inequality [137]. Let $\theta > 1$ and $R \le 2$ be two constants, and $0 \le \xi(x,t) \le 1$. Suppose η is an integrable nonnegative distribution solution of $\partial_t \eta - \mathscr{L}\eta + \xi(x,t)\eta = 0$ in $Q(\theta,R)$. Suppose $\|V\|_{L^\infty(Q(\theta,R))} \le 1$. Then there exists a constant $K_o > 0$ depending only on θ and the dimension such that

$$\inf_{|x| \le R/2} \eta(x,\theta R^2) \ge K_o \eta(0,R^2).$$

The inequality allows one to deduce the continuity (the relation between the minimum and maximum of a function at different points over a parabolically scaled domain Q). We apply it to the logarithm of the KPP solution $\log u(x,t)$. The maximum principle ensures that $u \in (0,1)$ for all (x,t).

Define $\xi(x,t,\hat{\omega}) = f(u(x,t,\hat{\omega}))/u(x,t,\hat{\omega})$. The KPP equation (5.52) is written as

$$\partial_t u = \Delta u + V(x,t,\hat{\omega}) \cdot \nabla u + \xi(x,t,\hat{\omega})u, \qquad (5.125)$$

where $\xi(x,t,\hat{\omega}) \in [0, f'(0)]$ is ready for the Harnack analysis.

The result is that if $\gamma(t) \ge 0$ is any nondecreasing function satisfying

$$\limsup_{t \to \infty} \frac{\gamma(t)}{t} \le \varepsilon,$$

then for any $c \in \mathbb{R}^d$,

$$\liminf_{t \to \infty} \frac{1}{t} \left(\log \inf_{|z| \le \gamma(t)} u(ct+z,t) - \log \sup_{y \in B_\delta(c(t-\gamma(t)))} u(y,t-\gamma(t)) \right)$$
$$\ge -C(1+|c|+\delta)^2 \varepsilon(1+\bar{V}_2) \qquad (5.126)$$

and

$$\limsup_{t \to \infty} \frac{1}{t} \left(\log \sup_{|z| \le \gamma(t)} u(ct+z,t) - \log \inf_{y \in B_\delta(c(t+\gamma(t)))} u(y, t+\gamma(t)) \right)$$
$$\le C(1+|c|+\delta)^2 \varepsilon (1+\bar{V}_2) \tag{5.127}$$

with probability one. Here, $\bar{V}_2$ is defined in (5.115) and $C = C(\theta)$ is a constant. Inequalities (5.126)–(5.127) lead to the tail estimates of the linearized solution. For $\delta > 0$, $x \in \mathbb{R}^d$, and $t \ge s \ge 0$, let $\phi(y,t;x,s) = \phi(y,t;x,s,\hat{\omega})$ satisfy the linear part of the KPP equation

$$\partial_t \phi = \Delta_y \phi + V \cdot \nabla \phi \tag{5.128}$$

for $t > s$ with the initial condition

$$\phi(y,s;x,s,\hat{\omega}) = \begin{cases} 1 & \text{if } y \in B_\delta(x), \\ 0 & \text{otherwise,} \end{cases}$$

at time $t = s$, where $\delta > 0$ is a fixed parameter. Define the family of functions

$$\phi^-(y,t;x,s) = \inf_{y' \in B_\delta(y)} \phi(y',t;x,s). \tag{5.129}$$

It follows from the maximum principle that for any $x,y,z \in \mathbb{R}^d$ and $r < s < t$, we have

$$\phi^-(z,t;x,r) \ge \phi^-(y,s;x,r)\phi^-(z,t;y,s). \tag{5.130}$$

For $c \in \mathbb{R}^d$ fixed, define the random process $q_{m,n}(\hat{\omega}) = \log \phi^-(cm,m;cn,n,\hat{\omega})$ indexed by integers m,n, $0 \le m < n$. We observe that $q_{m,n}$ is stationary and superadditive:

$$q_{m,n} \ge q_{m,k} + q_{k,n}, \quad \forall m < k < n,$$
$$q_{m+r,n+r}(\hat{\omega}) = q_{m,n}(\tau_{(cr,r)}\hat{\omega}). \tag{5.131}$$

One can show that $E[|q_{0,n}|] < \infty$ for all n. Then the subadditive ergodic theorem applies to $-q_{m,n}$, and we have that

$$-S(c) \stackrel{\Delta}{=} \lim_{n \to \infty} \frac{1}{n} q_{0,n} = \sup_{n > 0} \frac{1}{n} q_{0,n} \le 0 \tag{5.132}$$

exists almost surely and is nonrandom. The convexity of S follows from the subadditivity as shown in Chapter 4. By continuity estimates (5.126)–(5.127), the infimum in (5.129) may be replaced by the supremum for an error under control. More precisely, if $\gamma(t) \ge 0$ is any nondecreasing function satisfying $\limsup_{t \to \infty} \gamma(t)/t \le \varepsilon$, then for any $c \in Q^d$ (d- dimensional rational vectors),

$$\limsup_{t \to \infty} \frac{1}{t} \log \sup_{|z| \le \gamma(t)} \phi(ct+z,t;0,0) \le C(1+|c|+\delta)^2 \varepsilon (1+\bar{V}_2) - S(c) \tag{5.133}$$

and

$$\liminf_{t\to\infty} \frac{1}{t}\log \inf_{|z|\leq\gamma(t)} \phi(ct+z,t;0,0) \geq -C(1+|c|+\delta)^2\varepsilon(1+\bar{V}_2) - S(c) \quad (5.134)$$

hold with probability one. The large-deviation principle of ϕ follows, that for any open set $G \subset \mathbb{R}^d$,

$$\liminf_{t\to\infty} \frac{1}{t}\log \inf_{z\in tG} \phi(z,t;0,0,\hat{\omega}) \geq - \inf_{c\in G^o} S(c), \quad (5.135)$$

and for any closed set $F \subset \mathbb{R}^d$,

$$\limsup_{t\to\infty} \frac{1}{t}\log \sup_{z\in tF} \phi(z,t;0,0,\hat{\omega}) \leq - \inf_{c\in\bar{F}} S(c), \quad (5.136)$$

with probability one. The upper bound of the solution follows from the large-deviation upper bound (5.136) and the comparison inequality

$$u(y,t) \leq e^{tf'(0)}\phi(y,t;0,0),$$

by containing the support of initial data in a ball $B_\delta(0)$. So

$$\lim_{t\to\infty} \sup_{c\in F} u(ct,t) = 0,$$

for any compact set of c such that $S(c) > f'(0)$. As before, to show convergence of the KPP solution to one, we prove that the lower bound

$$\liminf_{t\to\infty} \frac{1}{t}\log \inf_{c\in K} u(ct,t) \geq - \max_{c\in K}(S(c) - f'(0)) \quad (5.137)$$

holds with probability one, for any compact set $K \subset \{c \in \mathbb{R}^d \mid S(c) - f'(0) > 0\}$. This is done by using $\phi(y,t;0,0)$ and its localizations as subsolutions, then extracting logarithmic asymptotics based on the large-deviation lower bound (5.135). The lower bound (5.137) and further construction of comparison functions of KPP solutions leads to the convergence of $u(ct,t)$ to one if $S(c) < f'(0)$.

5.4.2 Speed Bounds and Asymptotics

By working with PDE characterization (5.121) of the Lyapunov exponent μ and the variational formula (5.120) of c^*, we obtain the lower and upper bounds of c^* in terms of statistics of V and the front speed c_0 in the absence of advection. The PDE formulation (Eulerian method) is convenient for handling the divergence-free condition of V.

Proposition 5.21. *Suppose V is divergence-free and of mean zero: $E[V^{(j)}] = 0$ for $j = 1,\ldots,d$. The front speed c^* satisfies the upper bound*

$$c^*(e) \le c_0 + E_{\hat{p}}[\|V\|_{L_x^\infty}],$$

implying at most linear growth in $\delta \gg 1$ if V is scaled according to $V \mapsto \delta V$.
If $V(x,t)$ is uniformly bounded, then c^* also satisfies the lower bound

$$c^*(e) \ge c_0.$$

The above proposition extends similar bounds in the time-random shear flows as well as bounds for deterministic periodic flows. Numerical computation of c^* in randomly perturbed cellular flows [180] suggests that $c^* \sim O(\delta^p)$ at large δ may occur for any exponent $p \in (0,1)$, when V is scaled according to $V \mapsto \delta V$. So the above bounds are optimal in time-random incompressible flows.

The other type of bound on c^* for δV with Gaussian statistics in time is (iv) of Theorem 5.9, namely $c^* \le c_0\sqrt{1+\delta^2 p_1}$, where p_1 is the integral of the correlation function. We next give an extension of such bounds for the nonshear space–time-random flows.

The key assumption is that V is white in time. The following calculation is formal but illustrative. It uses the Lagrangian method (Feynman–Kac formula) to involve the second-order statistics of V, a change-of-measure Girsanov formula, and the properties of Wiener processes. A velocity field that is white noise in time could be incorporated rigorously through a term of the form $V \cdot \nabla u \circ dW$ in the original equation (5.52), where $\circ$ denotes the Stratonovich integral [125]. Although this scenario does not fall within our assumptions on V, the following derivation demonstrates the difficulty in estimating c^* when the velocity V is correlated in time.

Proposition 5.22. *Suppose that V has the form*

$$V(x,t,\hat{\omega}) = \sum_k X_k(x)F_k(t,\hat{\omega}), \qquad (5.138)$$

where $\{X_k(x)\}$ are periodic or almost periodic divergence-free fields and $\{F_k\}$ are white-noise processes in time, so that the covariance matrix function is

$$\Gamma_{ij} = \Gamma_{ij}(x_1, x_2, t_1 - t_2) = E_{\hat{p}}[V^{(i)}(x_1, t_1)V^{(j)}(x_2, t_2)] \le p_1 \delta_0(t_1 - t_2)A_{ij}(x_1, x_2),$$

where δ_0 is the standard delta function centered at zero, and p_1 is a constant. Assume that the KPP front speeds are given by (5.120). Then $c^* \le c_0\sqrt{1+C_2 p_1}$, where C_2 depends only on the dimension d and $f'(0)$.

Proof. The Feynman–Kac formula for φ^* of equation (5.121) gives

$$\varphi^*(x,0) = E\left[e^{-\lambda \cdot \int_0^t V(Z^\lambda, s)\, ds}\right] e^{|\lambda|^2 t},$$

where Z^λ is the diffusion process obeying the Itō equation

$$dZ^\lambda(s) = (V(Z^\lambda, s) - 2\lambda)\, ds + \sqrt{2}\, dW(s), \quad s \in [0,t],$$

$Z^{\lambda}(0) = z$, $W(s) = \{W^i(s)\}_{i=1}^d$ a d-dimensional Wiener process. Changing measure by the Girsanov theorem [125, Theorem 5.1] yields the following representation of φ^*:

$$E\left[\exp\left\{-\lambda\sqrt{2}\cdot W(t) + \sqrt{2}\sum_{i=1}^d \int_0^t V^{(i)}(W_z(r),r)\,dW^{(i)}(r)\right.\right.$$
$$\left.\left. -\frac{1}{2}\int_0^t \|V(W_z(s),s)\|^2\,ds\right\}\right], \qquad (5.139)$$

where $W_z(s) = z + W(s)$ and E is expectation with respect to W. It follows that

$$\varphi^* \le E\left[\exp\left\{-\lambda\sqrt{2}\cdot W(t) + \sqrt{2}\sum_{i=1}^d \int_0^t V^{(i)}(W_z(r),r)\,dW^{(i)}(r)\right\}\right]$$

and

$$E_{\hat{p}}\varphi^* \le E\left[e^{-\lambda\sqrt{2}\cdot W(t)}E_{\hat{p}}\left[\exp\left\{\sqrt{2}\sum_{i=1}^d \int_0^t V^{(i)}(W_z(r),r)\,dW^{(i)}(r)\right\}\right]\right]. \qquad (5.140)$$

Notice that inside the inner expectation (with $W_z(r)$ fixed), the sum of stochastic integrals is a linear combination of Gaussian variables. In other words, the inner expectation is over a log-normal variable, and so

$$E_{\hat{p}}\varphi^* \le E\left[\exp\left\{-\lambda\sqrt{2}\cdot W(t)\right.\right.$$
$$\left.\left. +\int_0^t\int_0^t \sum_{ij}\Gamma_{ij}(W(s),W(\tau),s,\tau)\,dW^{(i)}(s)\,dW^{(j)}(\tau)\right\}\right]. \qquad (5.141)$$

Since V is white in time, e.g., $\Gamma_{ij} = A_{ij}(x_1,x_2)p_1\delta_0(t_1-t_2)$, the integral in (5.141) is bounded from above by $p_1C_1\int_0^t \|dW(s)\|^2$. The right-hand-side expectation of (5.141) is bounded from above by

$$E\left[\exp\left\{\int_0^t p_1C_1\|dW(s)\|^2 - \sqrt{2}\lambda\cdot dW(s)\right\}\right]$$
$$= \prod_{j=1}^N\prod_{l=1}^d E\left[\exp\left\{p_1C_1\left(dW^{(l)}(s)\right)^2 - \sqrt{2}\lambda^{(l)}dW^{(l)}\right\}\right], \qquad (5.142)$$

where $dW^{(l)}$ is the Wiener increment over an interval of length t/N. We have used independence of Wiener increments in each component and among components. The last expression of (5.142) can be calculated explicitly, and equals, on taking the limit $N \to \infty$,

$$\exp\{|\lambda|^2 t + p_1 dC_1 t\}.$$

It follows that

$$\mu = \lim_{t \to \infty} \frac{1}{t} E_{\hat{p}} \log \varphi^* \le \lim_{t \to \infty} \frac{1}{t} \log E_{\hat{p}} \varphi^* \le |\lambda|^2 + C_1 d p_1, \tag{5.143}$$

or

$$c^* \le 2\sqrt{f'(0) + C_1 d p_1} = c_0 \sqrt{1 + C_2 p_1}. \tag{5.144}$$

This completes the proof.   □

Inequality (5.144) implies that rapid temporal decorrelation can reduce speed enhancement, the so-called speed-bending phenomenon. We remark that the physical mechanisms contributing to the speed-bending in a combustion process may be much more complicated and depend on whether the process is in gaseous or liquid phase and on activation energy, among others. Here we have identified time decorrelation as one mechanism for KPP fronts.

If V is Gaussian but nonwhite in time, then $p_1 \delta_0$ in the upper bound of the covariance matrix function is replaced by a nonnegative L^1 function with integral equal to p_1. The estimate of the right-hand-side expectation of (5.141) will be more complicated. It will be interesting to obtain a similar result as in the random-in-time shear flow case.

5.5 Stochastic Homogenization of Viscous HJ Equations

The study of KPP front speeds in Section 5.4 led to the homogenization of viscous HJs with quadratic Hamiltonian (5.123). A more general homogenization problem is to analyze the $\varepsilon \downarrow 0^+$ limit of the solutions of

$$u_t^\varepsilon = \varepsilon \sigma^2 \Delta u^\varepsilon + H(x/\varepsilon, t/\varepsilon, \omega, \nabla u^\varepsilon), \quad (t,x) \in (0, \infty) \times \mathbb{R}^d, \tag{5.145}$$

where $\sigma > 0$ is a constant (viscosity), $u^\varepsilon(x, 0) = g(x)$, with $g(x)$ a uniformly Lipschitz continuous function. The Hamiltonian $H = H(x, t, \omega, p)$ is a stationary ergodic process in (x, t) and is convex in p. Convergence of (5.145) is proved under structural assumptions of H in [135, 136]. In [150], the spatially random case is treated, allowing a degenerate viscous term and using a different method. Both results will be briefly discussed below.

To illustrate ideas, let us consider the spatially random case $H = H(x, \omega, p)$. The main assumptions on H are quite similar to those in Section 4.1 on inviscid HJs:

(H0) Convexity, stationarity, and ergodicity (see (A1)–(A2) in Section 4.1); the function $H(x, \omega, p)$ is equal to $H(p, \tau_x \omega)$, where τ_x is translation by x;

(H1) Coercivity: there exist constants $1 < \alpha \le \beta$, $c_1, c_2 > 0$ such that

$$c_1(|p|^\alpha - 1) \le H(p, \omega) \le c_2(|p|^\beta + 1)$$

holds for all (p, ω). The Legendre transform is well defined and gives a similar Lagrangian $L(q, \omega) = \sup_p [p \cdot q - H(p, \omega)]$.

(H2) Uniform continuity of $H(p, \tau_x \omega)$ in x, which implies that of $L(q, \tau_x \omega)$.

It follows from convexity that u^ε has a variational representation. Let $\varepsilon = 1$, and let $\mathscr{C}$ be the set of all bounded maps $c : [0, \infty) \times \mathbb{R}^d \to \mathbb{R}^d$. Consider the diffusion process (Itô equation)

$$x(t) = x + \int_0^t c(s, x(s)) \, ds + \sqrt{2} \sigma W(t), \tag{5.146}$$

where W is the standard Wiener process. Denote by Q_x^c the corresponding probability measure on the continuous sample path space of $x(t)$. Then

$$u(x, t, \omega) = \sup_{c \in \mathscr{C}} E^{Q_x^c} \left(g(x(t)) - \int_0^t L(x(s), \omega, c(s, x(s))) \, ds \right). \tag{5.147}$$

Rescaling time $t \to t/\varepsilon$, the scaled solution is ($\tilde{x} = x/\varepsilon$)

$$u^\varepsilon(x, t, \omega) = \sup_{c \in \mathscr{C}} E^c_{Q_{\tilde{x}}} \left(g(\varepsilon x(t/\varepsilon)) - \varepsilon \int_0^t L(x(s), \omega, c(s, x(s))) \, ds \right). \tag{5.148}$$

The next step is to pass to $\varepsilon \downarrow 0$ in the variational formula (5.148) and recover the Hopf–Lax formula in the limit with the help of stationarity, ergodicity, and continuity. By stationarity, it suffices to consider the limit of solutions at $x = 0$. Because $\sigma > 0$, the diffusion process (5.146) has an invariant measure for a special class of c and permits averaging by ergodic theory. When restricting the control c and performing averaging, one gets a lower bound of $\liminf u^\varepsilon(0, t)$. Such a class of control functions is in the form $c = b(\tau_x \omega)$, for some bounded function b. For given b, the probability associated with the diffusion process (5.146) has a density Φ, which is a solution in the distribution sense of the equation

$$\nabla \cdot (b\Phi) = \sigma^2 \Delta \Phi. \tag{5.149}$$

The averaging with respect to b and $x(t)$ occurs over large time, and it follows from (5.146) that almost surely in ω,

$$\lim_{\varepsilon \downarrow 0} \varepsilon x(t/\varepsilon) = \lim_{\varepsilon \downarrow 0} \varepsilon \int_0^{t/\varepsilon} b(\tau_{x(s)} \omega) \, ds = t E[b(\omega) \Phi(\omega)] \equiv tm(b, \Phi) \tag{5.150}$$

and

$$\lim_{\varepsilon \downarrow 0} \varepsilon \int_0^{t/\varepsilon} L(b(\tau_{x(s)} \omega), \tau_{x(s)} \omega) \, ds = t E[L(b(\omega), \omega) \Phi(\omega)] \equiv th(b, \Phi), \tag{5.151}$$

where expectation is over $\omega \in \Omega$.

Then the variational formula (5.148) gives the lower bound

$$\liminf u^\varepsilon(0, t) \geq \sup_{(b, \Phi) \in T_e} (g(tm(b, \Phi)) - th(b, \Phi)) = \sup_{y \in \mathbb{R}^d} (g(y) - t\bar{L}(y/t)), \tag{5.152}$$

where T_e denotes the space of pairs (b, Φ) satisfying the equation (5.149) and

$$\bar{L} = \bar{L}(q) = \inf_{\substack{(b,\Phi)\in T_e \\ E[b\Phi]=q}} h(b, \Phi) = \inf_{\substack{(b,\Phi)\in T_e \\ E[b\Phi]=q}} E[L(b(\omega),\omega)\Phi(\omega)]. \qquad (5.153)$$

We leave as an exercise to show that $\bar{L}$ is convex in q. The lower bound in (5.152) is the Hopf–Lax formula of the homogenized HJ with Hamiltonian $\bar{H}$, which is the Legendre transform of $\bar{L}$:

$$\bar{H} = \sup_{q\in R^d} (p\cdot q - \bar{L}) = \sup_{(b,\Phi)\in T_e} E[p\cdot b(\omega) - L(b(\omega),\omega)\Phi(\omega)]. \qquad (5.154)$$

The formula (5.154) is further used to find an upper bound of u^ε. It is first put into a dual variational form in which the constraint (5.149) is replaced by an additional layer of minimization

$$\bar{H} = \sup_\phi \sup_b \inf_\psi E\left[[p\cdot b(\omega) - L(b(\omega),\omega) + A_b\psi(\omega)]\phi(\omega)\right], \qquad (5.155)$$

where A_b is the adjoint operator of the one in the constraint (5.149), $A_b = \sigma^2\Delta + b(\omega)\cdot\nabla$. Formula (5.155) is easy to see by noting that for any ϕ,

$$\inf_\psi E[\phi A_b\psi(\omega)] = -\infty,$$

unless $(b, \psi) \in T_e$, which is preferred by $\sup_b$, since the corresponding value of $\inf_\psi$ is zero. The triple-layer min–max optimization in (5.155) is convenient for expressing $\bar{H}$ in terms of H:

$$\begin{aligned}
\bar{H} &= \sup_\phi \inf_\psi \sup_b E\left[[p\cdot b(\omega) - L(b(\omega),\omega) + A_b\psi(\omega)]\phi(\omega)\right] \\
&= \sup_\phi \inf_\psi \sup_b E\left[[(p+\nabla\psi)\cdot b(\omega) - L(b(\omega),\omega) + \sigma^2\Delta\psi]\phi(\omega)\right] \\
&= \sup_\phi \inf_\psi E\left[[H(p+\nabla\psi(\omega),\omega) + \sigma^2\Delta\psi]\phi(\omega)\right] \\
&= \inf_\psi \sup_\phi E\left[[H(p+\nabla\psi(\omega),\omega) + \sigma^2\Delta\psi]\phi(\omega)\right] \\
&= \inf_\psi \operatorname{ess\,sup}_\omega [H(p+\nabla\psi(\omega),\omega) + \sigma^2\Delta\psi], \qquad (5.156)
\end{aligned}$$

where the last equality follows from $0 \leq \phi \in L^1(\Omega)$. Hence for any small number $\delta > 0$, there is a function ψ_δ such that

$$H(p+\nabla\psi_\delta(\omega),\omega) + \sigma^2\Delta\psi_\delta \leq \bar{H} + \delta. \qquad (5.157)$$

An upper solution is

$$\bar{u}^\varepsilon(x,t) = p\cdot x + t(\bar{H}+\delta) + \varepsilon\psi_\delta(x/\varepsilon,\omega), \qquad (5.158)$$

for the essentially linear initial condition $u^\varepsilon(x,0) = p \cdot x + \varepsilon \psi_\delta(x/\varepsilon, \omega)$. Additional work [135] shows that $\varepsilon \psi_\delta(x/\varepsilon, \omega) \to 0$ for bounded x. It follows from (5.158) that

$$\limsup_{\varepsilon \to 0} u^\varepsilon(x,t) \le p \cdot x + t\bar{H},$$

removing arbitrarily small $\delta > 0$. An extension is possible for more general initial data under certain conditions of the Hamiltonian H [135, Sections 6 and 7]). More precisely, a main convergence result is given by the following theorem [135]:

Theorem 5.23. *Let* $u(x,t) = \sup_y (g(y) - t\bar{L}((y-x)/t))$. *Assume that (H0)–(H2) hold and that (H3) there exists a function* $v(\delta)$, $v \to 0$ *as* $\delta \to 0$, *such that for* $|x| \le \delta$, $\forall \omega \in \Omega$ *and a positive constant C:*

$$H(p,x,\omega) \ge (1+v(\delta))H((1+v(\delta))^{-1}p,\omega) - Cv(\delta).$$

Then with probability 1,

$$\lim_{\varepsilon \to 0} |u^\varepsilon(x,t,\omega) - u(x,t)| = 0,$$

uniformly in $(x,t) \in R^d \times [0,\infty)$.

The method above based on the ergodic theorem and min–max formulas developed in [135] extends to space–time random media [136] under slightly modified conditions (H0)–(H3), where spatial translation τ_x is replaced by space–time translation $\tau_{x,t}$ in (H2), and boundedness of x is replaced by that in (x,t). The space T_e is modified to

$$T_e = \{(b,\Phi) : \Phi_t + \nabla \cdot (b\Phi) = \sigma^2 \Delta \Phi\},$$

where derivatives are generated by the translation $\tau_{x,t}$ on Ω and Φ is a probability density. The homogenized Lagrangian $\bar{L}$ is given by the same formula (5.153), and the homogenized Hamiltonian $\bar{H}$ by (5.154). Under similar conditions of H, the homogenization of inviscid ($\sigma = 0$) HJ in space–time ergodic random media is proved in [213] by combining a subadditive ergodic theorem (Section 4.2) and continuity estimates.

Another approach to homogenization of spatially random viscous HJs is based on a subadditive ergodic theorem and uniform gradient estimates [150]. The subadditive ergodic theorem applies to H that grows faster than quadratically in p. When H has slower growth but is still coercive in the sense that $H \to \infty$ uniformly as $|p| \to \infty$, then H is perturbed (penalized) to $H + \eta|p|^m$, for $m > 2$ and a small $\eta > 0$. The solutions of the perturbed HJ equation are homogenized first, then one passes to the limit $\eta \to 0$. It is shown in [150] that such a double limit of penalized solutions agrees with the homogenization limit of the unperturbed HJ solutions ($\eta = 0$). The approach covers a wider class of HJ equations with degenerate viscosity of the form

$$u_t^\varepsilon - \varepsilon \operatorname{tr}(A(x,x/\varepsilon,\omega)\nabla^2 u^\varepsilon) + H(\nabla u^\varepsilon, u^\varepsilon, x, x/\varepsilon, \omega) = 0, \tag{5.159}$$

with bounded uniformly continuous initial datum u_0. Here, A and H are stationary and ergodic in y and ω, and A is a nonnegative matrix such that for any vector $\xi \in \mathbb{R}^d$, there is a positive constant Λ such that

$$0 \leq (A\xi, \xi) \leq \Lambda |\xi|^2,$$

where $(\cdot, \cdot)$ denotes inner product in $\mathbb{R}^d$. The key assumptions on H are as follows:

(A1) $H = H(p, r, x, y, \omega)$ is convex in p;

(A2) H is coercive in that:
$$H(p, r, x, y, \omega) \to +\infty,$$

uniformly in r, x, y, ω as $|p| \to +\infty$.

The main result is the following [150]:

Theorem 5.24. *Under (A1)–(A2), stationarity and ergodicity of A and H, and nonnegativity and boundedness of A, the solution u^ε of the HJ equation (5.159) converges to $\bar{u}$ in $C(\mathbb{R}^d \times [0, T])$ almost surely in ω. Here $\bar{u}$ satisfies the homogenized (deterministic) equation*

$$\bar{u}_t + \bar{H}(\nabla\bar{u}, \bar{u}, x) = 0, \quad (x, t) \in \mathbb{R}^d \times (0, T],$$

with the same initial datum u_0. The homogenized Hamiltonian $\bar{H}$ is convex.

We refer to [150] for details of the proof, including various approximations and error estimates.

Though the above homogenization results treat general convex nonlinearity or degenerate viscosity (diffusion), they do not allow unbounded random media, as seen from the coercivity condition (H1) or (A2). The methods in Sections 5.2 and 5.3 allow unbounded random media, yet for quadratic Hamiltonians. Homogenization of general convex viscous HJs with unbounded space–time randomness (without coercivity) remains an interesting issue for further research.

5.6 Generalized Fronts, Reactive Systems, and Geometric Models

In this section we discuss random fronts in non-KPP equations, systems of equations, and geometric models. The aim is to introduce a number of unsolved problems for future research based on recent progress.

5.6.1 Generalized Front Speeds and Central Limit Theorem

When reactive nonlinearity is non-KPP, one cannot find front speeds by analyzing solutions near zero. Instead, an estimate and control of the entire transition from zero to one must be available. This then relies on the dynamical properties of solutions. The KPP fronts are called pulled fronts (pulled by the unstable state $u = 0$), while the non-KPP fronts are pushed fronts (both regions where $u \approx 0$ and $u \approx 1$ contribute); see [210] for an overview of related physical literature. The notion of generalized transition front (GTF) in heterogeneous media has been recently proposed and studied for non-KPP reactions [159, 218, 21, 23, 162, 163, 174, 172], extending the known constant-speed traveling fronts. We shall give a definition of GTF below, in the context of bistable and ignition fronts (type 3 and type 5) in spatial random media.

Consider the scalar RD equation

$$u_t = u_{xx} + f(x,u), \tag{5.160}$$

where the spatial variation of f is arbitrary (aperiodic, nonergodic), and the initial condition $u(x,0)$ is a profile connecting one (at $-\infty$) to zero (at $+\infty$). A GTF is a global solution $u(x,t)$ for all $t \in \mathbb{R}$ such that $0 < u < 1$, and there is a continuous function (an interface) $X(t)$ such that for any $\varepsilon > 0$, there is a finite distance N_ε independent of t such that for all $t \in \mathbb{R}$,

$$u(x,t) > 1 - \varepsilon, \quad \forall x < X(t) - N_\varepsilon; u(x,t) < \varepsilon, \forall x > X(t) + N_\varepsilon. \tag{5.161}$$

This definition is adapted from the more general one in [21] and is reminiscent of the front-probing asymptotics in Chapters 2 and 3. It implies a finite width and speed of a frontlike global solution. Alternative definitions [159, 218] require a global-in-time solution and a certain continuity and invariance of its shape.

A solution $u = U(x,t,\omega)$ is called a random traveling wave [218] if it is a global solution such that $0 < U < 1$, $\lim_{x \to +\infty} U(x,0,\omega) = 0$, $\lim_{x \to -\infty} U(x,0,\omega) = 1$, and there exists a function $X(t,\omega)$ such that

$$U(x,t,\omega) = U(x - X(t,\omega), 0, \tau_{X(t,\omega)}\omega), \tag{5.162}$$

where τ is the spatial translation. The function $U(x,0,\omega)$ generates the random traveling wave, and plays the role of traveling-front profile in the case of periodic media in Chapter 2, where $X(t,\omega)$ also reduces to ct.

Now let us turn to equation (5.160) with spatially random reaction. Suppose that $f(x,u) = g(x,\omega)f_0(u)$, where $g(x,\omega)$ is a stationary ergodic process, almost surely uniformly Lipschitz in x, and is bounded by two deterministic constants

$$0 < g_{\min} \leq g(x,\omega) \leq g_{\max}.$$

For bistable f, one also requires that almost surely in ω,

$$\int_0^1 \inf_x (g(x,\omega)f_0(u))\, du \geq \varepsilon_b > 0,$$

for a positive constant $\varepsilon_b > 0$. Both conditions help to prevent front pinning by the random media. It is proved [174] that a GTF exists with the properties that $U_t > 0$, and there is a continuous increasing function $X(t)$, the interface, satisfying (1) $U(X(t),t) = \theta \in (0,1)$, (2) shape invariance (5.162) holds almost surely in ω, and (3)

$$\lim_{R \to +\infty} \sup_{t \in \mathbb{R}} \sup_{x > R} U(x + X(t), t, \omega) = 0,$$

$$\lim_{R \to +\infty} \inf_{t \in \mathbb{R}} \inf_{x < -R} U(x + X(t), t, \omega) = 1,$$

almost surely in ω. The GTF is proved [162] to be unique up to a constant translation in x and stable with perturbations decaying exponentially fast in time. Moreover,

$$\lim_{t \to \infty} X(t)/t = c^* \tag{5.163}$$

almost surely, where $c^* \in (c_{\min}, c_{\max}) \subset (0, +\infty)$ is the finite large-time front speed. The GTF satisfies all the definitions above. It has an invariant shape and finite width for all time, extending known traveling fronts in homogeneous and periodic media. A similar GTF is studied for a free boundary model in [45, 163].

Analogous to Burgers and HJ fronts in Chapters 3 and 4, the GTF speed fluctuations around c^*t is proved to be Gaussian under a condition of sufficient mixing of the random media [172]. See also [230] for a related study on the asymptotic tail behavior of a semilinear heat equation with a random source at the origin. The mixing condition means that the events related to the random media in the past ($x \sim -\infty$) and in the future ($x \sim +\infty$) are close to being independent; see Chapter 3 and [172] for precise mathematical characterizations. The main result is given in the following theorem [172]:

Theorem 5.25. *Consider the GTFs of equation (5.160) for bistable or ignition-type nonlinearity f. Assume that the random process $g(x, \omega)$ is sufficiently mixing. Then either (A) there is a positive constant $\kappa_0 > 0$ such that*

$$\lim_{t \to +\infty} P\left(\frac{X(t,\omega) - c^*t}{\kappa_0 \sqrt{t}} < \alpha \right) = N(\alpha), \quad \forall \alpha \in \mathbb{R}, \tag{5.164}$$

where

$$N(\alpha) = \frac{1}{\sqrt{2\pi}} \int_{-\infty}^c \exp\left\{ -\frac{y^2}{2} \right\} dy,$$

the unit Gaussian distribution function; or (B)

$$\lim_{t \to +\infty} P\left(\frac{X(t,\omega) - c^*t}{\sqrt{t}} > \alpha \right) = 0, \quad \forall \alpha \in \mathbb{R}. \tag{5.165}$$

In case (A), the invariance principle holds:

$$\frac{X(nt,\omega) - c^* nt}{\kappa_0 \sqrt{n}} \overset{\text{law}}{\longrightarrow} W(t), \quad t \in [0,1], \tag{5.166}$$

where $W(t)$ is the standard Wiener process. An example exists for case A.

The ideas of proofs for (5.163) and Theorem 5.25 are as follows. Let T_n be the random times at which the interface $X(t,\omega)$ reaches the integer points $x = n$. Since X is increasing in t, its inverse $T(x,\omega)$ is well defined such that $x = X(T(x,\omega),\omega)$, $T_n = T(n,\omega)$, $n = 1, 2, \ldots$. It follows from (5.162) that

$$U(x+y, T(y,\omega),\omega) = U(x,0,\tau_y\omega), \quad \forall y.$$

So as the front passes through the point y, the front profile is statistically invariant. Hence the sequence of increments $\Delta T_n = T_{n+1} - T_n$ is stationary and satisfies the law of large numbers (ergodic theorem [41, Section 6.5]):

$$\lim_{n \to +\infty} \frac{T_n}{n} = \lim_{n \to \infty} \frac{1}{n} \sum_{j=1}^{n-1} \Delta T_j = \frac{1}{c^*},$$

which is another form of (5.163). To prove (5.164)–(5.166), it suffices to demonstrate that

$$\lim_{n \to +\infty} E\left[\left|\frac{T_n - n/c^*}{\sqrt{n}}\right|^2\right] = \sigma^2, \tag{5.167}$$

for some constant $\sigma \geq 0$, and if $\sigma > 0$, then the family of processes

$$Z_n(x) = \frac{T_{xn} - nx/c^*}{\sigma\sqrt{n}} \overset{\text{law}}{\longrightarrow} W(x), \tag{5.168}$$

for $x \in [0,R]$ and any positive R. The difficulty is that the increments ΔT_n are correlated in a complicated way by the nonlinear PDE (5.160). What comes to the rescue is the stability property of the GTF, which implies that the interfacial motion depends primarily on the local environment, and only weakly on the distant past and future. In other words, the interfacial motion forgets its past and ΔT_n has enough decay of correlations at large times, and the Gaussian statistics of GTF speed fluctuation follows, except in a degenerate situation ($\sigma = 0$).

It is not known whether CLT is true for KPP fronts, though its quadratic viscous HJ approximation obeys CLT, as shown in Chapter 4. The limit law (5.163) is the non-KPP analogue of the KPP front speeds in a one-dimensional spatial random medium [94, 100], which served as a fundamental and inspiring first step in the study of RD fronts in random media. In the KPP case, (5.163) holds without the knowledge of GTF. In fact, GTF is not known to exist for KPP, type-2 and type-4 reactions. It remains to study GTF in random flows in multiple space dimensions and compare with KPP fronts.

5.6.2 Fronts in Reaction–Diffusion Systems

The RD systems in combustion or autocatalytic reactions [238, 28, 33, 53] are of
the following form:

$$u_t = \Delta_x u + v f(u), \quad x \in \mathbb{R}^d,$$
$$v_t = \mathrm{Le}^{-1}\Delta_x v - v f(u), \tag{5.169}$$

where $\mathrm{Le} > 0$ is called the Lewis number. If (5.169) models premixed flame fronts
in a one-step exothermic chemical reaction of the form $A \rightarrow B$, then u is the temperature of the reacting mixture and v is the mass fraction of the reactant A. The
function f takes the Arrhenius form $e^{-E/T}$, with activation energy constant $E > 0$.

If $d = 1$, then adding the two equations shows that $u + v$ satisfies the heat equation
and hence is forever equal to one if this is so arranged at $t = 0$. Replacing v by $1 - u$
in the first equation of (5.169), we find a scalar R-D equation of type 4, and type 5
then arises as we introduce a temperature cutoff θ.

For existence and uniqueness of traveling fronts of the form $(u,v) = (U(p \cdot x -
ct), V(p \cdot x - ct))$, p a given unit vector, see [28, 37, 52, 155]. It is well known that
if d is much larger or smaller than 1, then fronts are unstable; see [14, 26, 123, 223]
and references therein. Intuitively, the very distinct diffusion constants cause the
front to develop spatial–temporal scales as a way of keeping balance. With $\mathrm{Le} > 1$,
fronts oscillate in time, and with $\mathrm{Le} < 1$, they generate transverse spatial oscillations
in two or three dimensions. The scales continue to grow with Le, and eventually the
solutions are chaotic.

When (5.169) models isothermal autocatalytic reactions of the form $A + mB \rightarrow
(m+1)B$, $m \geq 1$, with rate law proportional to vu^m, where v and u are the concentra-
tions of the reactant A and the catalyst B, the function f is now $f(u) = vu^m$. Again
when $\mathrm{Le} = 1$, we recover a scalar R-D equation of type 1 if $m = 1$, of type 2 if $m \geq 2$.
Existence and dynamics of fronts are discussed in [33, 34, 35, 52, 53, 93]. Similarly,
if Le is sufficiently far away from one, fronts are unstable and can be chaotic; see
[119, 154, 164].

Scalar R-D equations of type 3 come from the FitzHugh–Nagumo (FHN) system
in mathematical biology,

$$u_t = \Delta_x u + u - u^3 - v, \quad x \in \mathbb{R},$$
$$v_t = \varepsilon(u - \gamma v), \tag{5.170}$$

where $\gamma > 0$, and $\varepsilon > 0$ is a small parameter. In the limit $\varepsilon \rightarrow 0$, (5.170) reduces to
a bistable scalar R-D equation; see [89, 122, 167, 170, 211].

The front problem of KPP systems in a shear flow was studied recently [111].
The system of equations is (T temperature, and Y concentration of reactant)

$$T_t + u(y)T_x = \Delta_{x,y}T + f(T)Y,$$
$$Y_t + u(y)Y_x = \mathrm{Le}^{-1}\Delta_{x,y}Y - f(T)Y, \tag{5.171}$$

where $(x,y) \in \mathbb{R} \times \Omega$, $u(y)$ is Hölder continuous, $\int_\Omega u(y)dy = 0$, and f is of KPP type: f is continuously differentiable, and

$$f(0) = 0 < f(s) \leq f'(0)s, \quad f'(s) \geq 0, \forall s > 0, \ f(+\infty) = +\infty.$$

In particular $f(T) = T$. The boundary conditions are zero Neumann on $\mathbb{R} \times \partial\Omega$. The traveling fronts are solutions of the form $T = \tilde{T}(x - ct, y)$, $Y = \tilde{Y}(x - ct, y)$, so that $(\tilde{T}, \tilde{Y})(+\infty, y) = (0,1)$, $(\tilde{T}, \tilde{Y})(-\infty, y) = (1,0)$, uniformly in $y \in \Omega$. The main finding is contained in the following theorem [111]:

Theorem 5.26. *For any* Le > 0, *the minimal KPP front speed c^* is same as that in the case of the unit Lewis number. If and only if $c \geq c^*$, a traveling front $T = \tilde{T}(x - ct, y)$, $Y = \tilde{Y}(x - ct, y)$ exists such that $T > 0$, $0 < Y < 1$, T is bounded.*

The interesting fact is that the KPP minimal speed is independent of Le > 0. This was first pointed out in [33] for homogeneous media. In [53], c^* is also proved to be the large-time asymptotic speed selected by compactly supported initial data of $T(x,0)$ $(Y(x,0) = 1)$ in the absence of flow.

It would be very interesting to confirm the Le independence of c^* for more general flows, especially random flows.

The existence of traveling fronts is studied in [22, 110] when the boundary conditions contains a heat loss, e.g., $\frac{\partial}{\partial n}T + \sigma T = 0$ on $\mathbb{R} \times \Omega$, n the unit normal direction, and the minimal speed $c^* = c^*(\sigma)$ obeys a variational principle. The proofs are based on topological degree theory and PDE estimates. When f is non-KPP, traveling fronts are studied in the perturbative regime in which Le is near one; see [69, 70] for the ignition nonlinearity.

5.6.3 Geometric Models and Huygens Fronts

One way to model motion of an interface (a curve in $\mathbb{R}^2$, a surface in $\mathbb{R}^3$) is to prescribe a motion law in terms of its normal velocity, denoted by V_n. The simplest law is that $V_n = c$, a constant. We shall represent the interface by a constant level set of a scalar function $G = G(x,t)$; see [187] and references therein for level set theory, numerical methods, and applications.

In the level set formulation, the normal velocity satisfies $V_n = -G_t/|\nabla_x G|$. For example, an expanding circle (sphere) can be represented as $G = |x|^2 - t^2 = 0$, $x \in \mathbb{R}^2$ ($\mathbb{R}^3$). The inside part is $G < 0$, and the outside part is $G > 0$. The derivatives of the level function are $G_t = -2t$, $\nabla G = 2x$, $|\nabla G| = 2|x|$. When restricted to the level $G = 0$, we have $|\nabla G| = 2|x| = 2t$, and so $-G_t/|\nabla_x G| = 1$ gives the unit normal velocity pointing from the inside ($G < 0$) to the outside ($G > 0$). The outward-pointing unit normal is $\mathbf{n} = \nabla G/|\nabla G|$, which is simply $x/|x|$ in our example. In general, we shall consider an expanding closed curve or surface, as shown in the experimental picture of Chapter 1. A geometric quantity is the mean curvature defined as the divergence of the normal,

$$\kappa = \nabla \cdot \mathbf{n} = \nabla \cdot \nabla G / |\nabla G|, \tag{5.172}$$

which is positive (negative) if the interface is locally convex (concave). The mean curvature is $\kappa = 1/|x|$ ($\kappa = 2/|x|$) for a circle (sphere) of finite radius.

The next-simplest law is $V_n = g(x,t)$, g being a given space–time function. In combustion, $g(x,t)$ is chosen as the sum of a constant laminar speed $-s_L$ and the local fluid velocity $\mathbf{v}(x,t)$ along the normal $\mathbf{n}$ of a thin flame [238]. So

$$V_n = -\frac{G_t}{|\nabla_x G|} = -s_L + \mathbf{v}(x,t) \cdot \mathbf{n} = -s_L + \frac{\mathbf{v} \cdot \nabla G}{|\nabla G|},$$

which is just the Hamilton–Jacobi equation

$$G_t + \mathbf{v}(x,t) \cdot \nabla G = s_L |\nabla G|, \tag{5.173}$$

known as the G-equation [156, 238].

The $G = 0$ isocontour represents implicitly the reaction surface of a moving flame front whose width is infinitesimal. Such fronts are also called Huygens fronts; their dynamics depend only on the local environment. The Hamiltonian function of (5.173) is $H = H(p,x,t) = s_L |p| - \mathbf{v}(x,t) \cdot p$. Recall the remark at the end of Section 2.4; the asymptotic HJ equation for bistable and ignition fronts has a linearly growing (relativistic) Hamiltonian $H(p,x,t) = O(|p|)$ at large $|p|$ for fixed (x,t), which is also the case for the G-equation (5.173). In contrast, the KPP Hamiltonian is quadratic (classical) in p, $H = O(|p|^2)$. From the Hamiltonian perspective, the G-equation models the bistable and ignition fronts better than the KPP fronts.

The G-equation has been widely adopted in the combustion literature. Various analytical and numerical approximations on front speeds in periodic and random flows are based on it; see [126, 127, 251, 224, 11, 204, 54, 194, 1, 2, 51], among many others. A comparative study of front speeds in the G-equation and the KPP equation is conducted in [76], where examples of shear flow with nonzero mean show that the front speeds from the G-equation are less than the KPP speeds. The amount of discrepancy may vary depending on the alignment of the mean flow and the shear flow. In [77], the validity of Huygens fronts and the G-equation is studied for piecewise linear reactions and linear incompressible flows. In spite of the differences between Huygens fronts and RD fronts, theoretical estimates [251] from the G-equation using a formal renormalization group (RG) method have been found to match the empirical speed growth laws for the aqueous autocatalytic chemical reaction fronts [222]. The flows generated in such experiments range from capillary-wave flow, Taylor–Couette flow, to vibrating-grid turbulence [222]. For the G-equation (5.173) with space–time random velocity field $\mathbf{v}$, the (turbulent) front speed s_T is defined as $s_L \langle |\nabla G| \rangle$, where the bracket is the ensemble average. The formula for s_T by RG analysis [251] is

$$U_T = \exp\{(U/U_T)^p\} \tag{5.174}$$

for $p = 2$, where $U_T = s_T / s_L$, and U is the ratio of the root-mean-square amplitude of $\mathbf{v}$ and s_L. For large U, (5.174) implies that U_T is approximately $U / \sqrt{\ln U}$.

In terms of establishing the almost sure existence of s_T from stochastic homogenization methods, one observes immediately that the G-equation violates the coercivity condition (H1) or (A2) or similar assumptions in Chapter 4. For a mean-zero-velocity field $\mathbf{v}$ with larger amplitude than s_L, we have that $H(p,x,t)$, although still convex in p, does not grow to $+\infty$ as $|p| \to +\infty$. The current mathematical theory does not apply. Stochastic homogenization of the G-equation remains an interesting topic for future research.

The G-equation (5.173) clearly ignores diffusion, as suggested by the viscous HJ of the KPP equation (5.123). A generalized G-equation has been proposed [193] to model fronts in gaseous combustion systems. Some physical length scales are introduced into (5.173). A characteristic length for flame is called the Markstein length $L_m = l_f \alpha(\text{Le}, E)$, where l_f is flame thickness, and α is a nondimensional number depending on the Lewis number Le and the activation energy E, [193, (1.13)]. The generalized motion law [193] is derived by replacing s_L in (5.173) by

$$\tilde{s}_L = s_L(1 + L_m \kappa) + L_m \mathbf{n} \cdot \nabla \mathbf{v} \cdot \mathbf{n}, \tag{5.175}$$

where the correction terms contribute to flame-stretching due to flame geometry and flows. The mean curvature κ is a second-order term similar to diffusion. In fact, it is equal to Δdist_f, where dist_f is the signed distance function to the front.

So the generalized G-equation is a viscous (parabolic) stochastic HJ equation worthy of analysis and qualitative comparison with (5.173) in terms of the properties of their (turbulent) front speeds s_T.

Geometric models have also been proposed for the study of phase boundary motion through a heterogeneous material (a matrix with precipitates) [60, 61]. The motion law is

$$V_n = f(x) - c\kappa,$$

which becomes, in the level set formulation,

$$h_t = -f(x)|\nabla h| + c|\nabla h|\nabla \cdot (\nabla h/|\nabla h|), \tag{5.176}$$

where $f(x)$ is the jump in energy density across the interface and c is a nonnegative constant. The term $f - c\kappa$ is a thermodynamic force driving the phase boundary [61]. If f is a constant plus noise, the model (5.176) also appears in statistical mechanics [192] and the study of dislocation loops in materials [108]. In case of dislocation, the function $f(x)$ is equal to a constant except on point defects or inclusions. The defects may be periodically or randomly distributed. If $f(x)$ is periodic and has a fixed sign (hence coercive), then periodic homogenization (scale c to εc and f to $f(x/\varepsilon)$) is studied in the spirit of [148, 79] for both $c = 0$ [60] and $c > 0$ [61]. In the case $c > 0$ and f does not change sign, the homogenization problem is treated in [151] for both periodic and almost-periodic media. The homogenized motion is $V_n = \bar{f}(\mathbf{n})$ for some continuous function $\bar{f}$, implying an anisotropic geometric law. See [60] for variational formulas of $\bar{f}$ in the case $c = 0$, and [61] for an explicit formula of $\bar{f}$ in the large-c limit. If f changes sign, front-trapping may occur [60, 61, 67]. The existence and uniqueness of pulsating traveling fronts with constant speeds are proved [66] if f changes sign and is small enough and if $c > 0$. The stochastic homogenization problem of (5.176) remains largely open.

5.7 Exercises

1. Show that the Lyapunov exponent $\mu(\lambda)$ at $(\lambda_1, 0)$, $\lambda_1 \in \mathbb{R}$, for the time-random shear flow $b = b(y, t, \omega)$ satisfies the inequality $\mu(\lambda_1, 0) \geq \lambda_1^2/2$. Then deduce from the speed variational formula that $c^*(\delta)$, the front speed in scaled shear flow $\delta b(y, t)$, is no less than $c^*(0)$, the speed in the absence of flow.

2. Make the change of variable $\psi = \log(\phi)$ in the equation (5.77) and derive a viscous quadratic HJ equation for ψ. Then show that $c^*(\delta) = c^*(0)$ if and only if $b = b(t)$.

3. Derive the linear upper bound (iii) of Theorem 5.9 by the Feynman–Kac formula
$$\phi = E\left[e^{\lambda \delta \int_0^t b(W(s), t-s)\, ds} \right].$$

4. Verify that if the shear flow b is a mean-zero Gaussian process, then for any fixed continuous path $W \in C([0,1], \mathbb{R}^2)$, the random variable
$$\xi(t, W) \equiv e^{-\int_0^t \lambda_1 b(W_2(s)+z, t-s)\, ds - \lambda_1 W_1(t) - \lambda_2 W_2(t)}$$
is lognormal with mean equal to
$$E[\xi(t, W)] = e^{|\lambda|^2 \hat{\sigma}^2/2} e^{-\lambda_1 W_1(t) - \lambda_2 W_2(t)},$$
where
$$\hat{\sigma}^2 = \int_0^t \int_0^t \Gamma(W(s), W(r), s, r)\, ds\, dr,$$
where Γ is the covariance function of the process b.

References

1. M. Abel, A. Celani, D. Vergni, A. Vulpiani, *Front propagation in laminar flows*, Phys. Review E 64, 046307 (2001), 1–12.
2. M. Abel, M. Cencini, D. Vergni, A. Vulpiani, *Front speed enhancement in cellular flows* 12(2) (2002), 481–488.
3. R. Abraham, J.E. Marsden, *Foundations of Mechanics*, 2nd edition, Addison-Wesley Publishing Co, Reading, MA, 1978.
4. R. Adler, *The Geometry of Random Fields*, John Wiley and Sons, 1980.
5. N. Alikakos, P. Bates, X. Chen, *Traveling waves in a time periodic structure and a singular perturbation problem*, Transactions of AMS 351 (1999), 2777–2805.
6. E. Anderson, et. al., *LAPACK Users' Guide*, Third Edition, Society for Industrial and Applied Mathematics, 1999.
7. V. Arnold, *Mathematical Method of Classical Mechanics*, Springer-Verlag, 1978.
8. D. G. Aronson and H. F. Weinberger, *Multidimensional nonlinear diffusion arising in population genetics*, Adv. in Math. 30 (1978), 33–76.
9. D. Aronson and H. Weinberger, *Nonlinear diffusion in population genetics, combustion, and nerve propagation*. Lecture Notes in Mathematics 446, pp. 5–49, Springer-Verlag, 1975.
10. W. Ashurst, *Flow-frequency effect upon Huygens front propagation*, Combust. Theory Modelling 4 (2000), 99–105.
11. W. Ashurst, G.I. Sivashinsky, *On flame propagation through periodic flow fields*, Combust. Sci. Technol. 80 (1991), 159.
12. B. Audoly, and H. Berestycki, and Y. Pomeau, *Réaction-diffusion en ecoulement stationnaire rapide*, Note C. R. Acad. Sci. Paris Sér. II 328 (2000), 255–262.
13. A.-L. Barabási and H. E. Stanley, *Fractal Concepts in Surface Growth*, Cambridge Univ. Press, Cambridge, 1995.
14. G. I. Barenblatt, Y. B. Zeldovich and A. G. Istranov, *On the diffusional-thermal stability of a laminar flame*, Zh. Prikl. Mekh. Tekh. Fiz. 4 (1962), 21–26.
15. J. Bebernes, C. Li, Y. Li, *Travelling fronts in cylinders and their stability*, Rocky Mountain J. of Math, 27(1) (1997), 23–150.
16. R. D. Benguria and M. C. Depassier, *Variational characterization of the speed of propagation of fronts for the nonlinear diffusion equation*, Comm. Math. Phys. 175 (1996), 221–227.
17. R. D. Benguria and M. C. Depassier, *Speed of fronts of the reaction-diffusion equation*, Phys. Rev. Lett. 77 (1996), 1171–1173.
18. A. Bensoussan, J.-L. Lions and G. Papanicolaou, *Asymptotic Analysis for Periodic Structures*. Studies in Mathematics and its Applications, Vol. 5, North-Holland, Publ., 1978.

19. H. Berestycki, *The influence of advection on the propagation of fronts in reaction-diffusion equations*, in Proceedings of the NATO ASI Conference, Cargese, France, H. Berestycki and Y. Pomeau, eds., Kluwer, Dordrecht, the Netherlands, 2003.

20. H. Berestycki, F. Hamel, *Front propagation in periodic excitable media*, Comm. Pure and Appl. Math, 55 (2002), 949–1032.

21. H. Berestycki, F. Hamel, *Generalized traveling waves for reaction-diffusion equations*, Contemp. Math 446, Amer. Math. Soc., pp. 101–123, 2007.

22. H. Berestycki, F. Hamel, A. Kiselev, L. Ryzhik, *Quenching and propagation in KPP reaction-diffusion equations with a heat loss*, Arch. Ration. Mech. Anal 178 (2005), 57–80.

23. H. Berestycki, F. Hamel, H. Matano, *Bistable travelling waves around an obstacle*, preprint, 2008.

24. H. Berestycki, F. Hamel, and N. Nadirashvili, *Elliptic eigenvalue problems with large drift and applications to nonlinear propagation phenomena*, Comm. Math. Phys, 253 (2005), 451–480.

25. H. Berestycki, F. Hamel, L. Roques, *Analysis of the periodically fragmented environment model: 2-biological invasions and pulsating travelling fronts*, J. Math. Pures Appl. 84 (2005), 1101–1146.

26. H. Berestycki and B. Larrouturou, *Quelque aspect mathématique de la propagation des flames prémélangée*, p. 65, Pitman Research Notes in Mathematics Series, H. Brezis and J.-L. Lions eds., Vol. X, 1991.

27. H. Berestycki, B. Larrouturou and P.-L. Lions, *Multi-dimensional travelling-wave solutions of a flame propagation model*. Arch. Rat. Mech. Anal. 111 (1990), 33–49.

28. H. Berestycki, B. Nicolaenco and B. Scheurer, *Traveling wave solution to combustion models and their singular limits*, SIAM J. Math. Anal. 16 (1985), 1207–1242.

29. H. Berestycki and L. Nirenberg, *Some qualitative properties of solutions of semilinear elliptic equations in cylindrical domains*, Analysis etc., ed., P. Rabinowitz, et al., Academic Pr., 1990, pp. 115–164.

30. H. Berestycki and L. Nirenberg, *On the method of moving planes and the sliding method*, Bol. da Soc. Brasiliera de Matematica, 22 (1991), 1–37.

31. H. Berestycki and L. Nirenberg, *Traveling Fronts in Cylinders*, Annales de l'IHP, Analyse non linéaire 9 (1992), 497–572.

32. J. Billingham and D. J. Needham, *A note on the properties of a family of traveling wave solutions arising in cubic autocatalysis*, Dynamics Stabil. Systems 6 (1991), 33–49.

33. J. Billingham and D. J. Needham, *The development of traveling waves in quadratic and cubic autocatalysis with unequal diffusion rates, I. Permanent form traveling waves*, Phil. Trans. R. Soc. Lond. A 334 (1991), 1–24.

34. J. Billingham and D. J. Needham, *The development of traveling waves in quadratic and cubic autocatalysis with unequal diffusion rates, II. The initial value problem with an immobilized or nearly immobilized autocatalyst.*, Phil. Trans. Roy. Soc. Lond. A 336 (1991), 497–539.

35. J. Billingham and D. J. Needham, *The development of traveling waves in quadratic and cubic autocatalysis with unequal diffusion rates, III. Large time development in quadratic autocatalysis*, Quart. Appl. Math. 1 (1992), 343–372.

36. P. Billingsley, *Convergence of Probability Measures*, Wiley, 1968.

37. A. Bonnet, B. Larrouturou and L. Sainsaulieu, *On the stability of multiple steady planar flames when the Lewis number is less than one*, Phys. D 69 (1993), 345–352.

38. W. Bosma and S. van der Zee, *Transport of reacting solute in a one-dimensional chemically heterogeneous porous media*, Water Resources Research 29 (1993), 117–131.

39. W. Bosma, S. van der Zee and C. J. van Duijn, *Plume development of a nonlinearly adsorbing solute in heterogeneous porous formations*, Water Resources Research 32 (1996), 1569–1584.

40. M. Bramson, *Convergence of solutions of the Kolmogorov equations to traveling waves*, Mem. of AMS 285 (1983).

41. L. Breiman, *Probability*, Classics in Applied Mathematics, SIAM, 1992.
42. J. Bricmont and A. Kupiainen, *Stability of moving fronts in the Ginzburg–Landau equations*, Comm. Math. Phys. 159 (1994), 287–318.
43. R. J. Briggs, *Electron-Stream Interaction with Plasmas*, MIT press, Cambridge, 1964.
44. N. F. Britton, *Reaction-Diffusion Equations and Their Applications to Biology*, London, Academic Press, 1986.
45. L. Caffarelli, A. Mellet, *Flame propagation in one-dimensional stationary ergodic media*, Math. Models Methods Appl Sci 17 (2007), 155–169.
46. L. Caffarelli, K. Lee, A. Mellet, *Singular limit and homogenization for flame propagation in periodic excitable media*, Arch. Rat. Mech. Analy. 172(2) (2004), 153–190.
47. L. Caffarelli, K. Lee, A. Mellet, *Homogenization and flame propagation in periodic excitable media: the asymptotic speed of propagation*, Comm. Pure Appl. Math 59(4) (2006), 501–525.
48. R. Carmona, S. Molchanov, *Parabolic Anderson problem and intermittency*. Mem. Amer. Math. Soc. 108 (1994), no. 518, viii+125.
49. R. Carmona, S. Molchanov, *Stationary parabolic Anderson model and intermittency*, Prob. Theory Rel. Fields 102 (1995), 433–453.
50. R. Carmona, L. Koralov, S. Molchanov, *Asymptotics for the almost sure Lyapunov exponent for the solution of the parabolic Anderson problem*, Random Oper. and Stoch. Equ. 9(1) (2001), 77–86.
51. M. Cencini, A. Torcini, D. Vergni, A. Vulpiani, *Thin front propagation in steady and unsteady cellular flows*, Phys. of Fluids 15(3) (2003), 670–688.
52. X. Chen, Y. Qi, *Sharp estimates on minimum traveling wave speed of reaction diffusion systems modeling autocatalyst*, SIAM J. Math Analysis 39 (2007), 437–448.
53. X. Chen, Y. Qi, *Propagation of local disturbances in reaction diffusion systems modeling quadratic autocatalysis*, SIAM J. Appl. Math 69(1) (2008), 273–282.
54. M. Chertkov, V. Yakhot, *Hyugens front propagation in turbulent fluids*, Phys. Rev. Lett 80 (1998), 2837–2841.
55. A. Chorin, O. Hald, *Stochastic Tools in Mathematics and Science*, Vol. 1, Surveys and Tutorials in the Applied Mathematical Sciences, Springer, 2006.
56. P. Clavin and F. A. Williams, *Theory of premixed-flame propagation in large-scale turbulence,* J. Fluid Mech. 90 (1979), 598–604.
57. M. Concordel, *Periodic homogenization of Hamilton-Jacobi equations I: additive eigenvalues and variational formula*, Indiana Univ math J 45 (1996), 1095–1117.
58. J. Conlon, C. Doering, *On Traveling Waves for the Stochastic FKKP Equation*, Jour. Stat Physics 120(3-4) (2005), 421–477.
59. P. Constantin, A. Kiselev, A. Oberman, and L. Ryzhik, *Bulk burning rate in passive-reactive diffusion*, Arch. Ration. Mech. Anal. 154 (2000), 53–91.
60. B. Craciun and K. Bhattacharya, *Homogenization of a Hamilton-Jacobi equation associated with the geometric motion of an interface*, Proceedings of the Royal Society Edinburgh A. 133 (2003), 773–805.
61. B. Craciun and K. Bhattacharya, *Effective motion of a curvature sensitive interface through a heterogeneous medium*, Inter. Free. Bound. 6 (2004), 151–173.
62. M. Cranston and T. Mountford, *Lyapunov exponent for the parabolic Anderson model in R^d*, Journal of Functional Analysis 236 (2006), 78–119.
63. G. Dee, J.S. Langer, *Propagating Pattern Selection*, Phys. Rev Lett 50 (1983), 383–387.
64. D. A. Dawson and E. Perkins, *Historical processes*, Mem. Amer. Math. Soc. 93 (1991).
65. B. Denet, *Possible role of temporal correlations in the bending of turbulent flame velocity*, Combust. Theory Modelling 3 (1999), 585–589.
66. N. Dirr, G. Karali, A. Yip, *Pulsating Wave for Mean Curvature Flow in Inhomogeneous Medium*, European J. of Applied Mathematics 19 (2008), 661–699.
67. N. Dirr, A. Yip, *Pinning and De-Pinning Phenomena in Front Propagation in Heterogeneous Medium*, Interfaces and Free Boundaries, 8 (2006), 79–109.

68. C. Doering, C. Mueller, P. Smereka, *Interacting particles, the stochastic FKPP equation and duality*, Phys. A 325 (2003), 243–259.

69. A. Ducrot, *Multi-dimensional combustion waves for Lewis number close to one*, Math. Methods Appl. Sci. 20 (2007), 291–304.

70. A. Ducrot, M. Marion, *Two-dimensional traveling wave solutions of a system modelling near equi-diffusional flames*, Nonlinear Anal 61 (2005), 1105–1134.

71. C. J. van Duijn and P. Knabner, *Solute transport in porous media with equilibrium and nonequilibrium multiple sites adsorption, travelling waves*, J. Reine Angew. Math. 415 (1991), 1–49.

72. R. Durrett, *Probability: Theory and Examples*, 2nd ed., Wadsworth and Brooks, 1995.

73. Weinan E, Y. Sinai, *New results in mathematical and statistical hydrodynamics*, Russian Math. Surveys 55(4) (2000), 635–666.

74. Weinan E, J. Wehr, J. Xin, *Breakdown of homogenization for the random Hamilton-Jacobi equations*, Comm. Math Sci, 6(1) (2008), 189–197.

75. J-P. Eckmann and C.E. Wayne, *The nonlinear stability of front solutions for parabolic pde's*, Comm. Math. Phys. 161 (1994), 323–334.

76. P. Embid, A. Majda and P. Souganidis, *Comparison of turbulent flame speeds from complete averaging and the G-equation*, Phys. Fluids 7(8) (1995), 2052–2060.

77. P. Embid, A. Majda and P. Souganidis, *Examples and Counterexamples for Huygens Principle in Premixed Combustion*, Combustion Sci and Tech. 120 (1996), 73–303.

78. L. C. Evans, *Weak Convergence Methods for Nonlinear PDE's*, NSF-CBMS Regional Conference Lectures, 1988.

79. L. C. Evans, *Periodic homogenization of certain fully nonlinear partial differential equations*, Proc. Royal Soc. Edingburgh, Section A 127 (1992), 65–689.

80. L. C. Evans, *Partial Differential Equations*, Graduate Studies in Mathematics, American Mathematical Society, Providence, R.I, 1998.

81. L. C. Evans and P. E. Souganidis, *A PDE approach to geometric optics for certain semilinear parabolic equations*, Indiana Math. J., 38 (1989), 141–172.

82. L. C. Evans and P. E. Souganidis, *A PDE approach to certain large deviation problems for systems of parabolic equations*, Ann. Inst. H. Poincaré, Anal. Nonlinéare, 6 (1989), Suppl., 229–258.

83. L. C. Evans and P. E. Souganidis, *Differential games and representation formulas for solutions of Hamilton–Jacobi–Isaacs equations*, Indiana Math. J. 33 (1984), 773–797.

84. E. B. Fabes and D. W. Stroock, *A new proof of Moser's parabolic Harnack inequality using the old ideas of Nash*, Arch. Rat. Mech. Anal. 96 (1986), 327–338.

85. H. Fan, *Large time behavior of elementary waves of Burgers equation under white noise perturbations*, Comm. Partial Differential Equations, 20(9-10) (1995), 1699–1723.

86. A. Fannjiang and G. Papanicolaou, *Convection enhanced diffusion for periodic flows*, SIAM J. Appl. Math, 54 (1994), 333–408.

87. A. Fannjiang and G. Papanicolaou, *Convection enhanced diffusion for random flows*, Journal of Statistical Physics 88 (1997), 1033–1076.

88. G. Fennemore and J. Xin, *Wetting fronts in one dimensional periodically layered soils*, SIAM J. Applied Math. 58 (1998), 387–427.

89. P. C. Fife, *Mathematical Aspects of Reacting and Diffusing Systems*, Lecture Notes in Biomath, 28, Springer, New York, 1979.

90. P. C. Fife, *Dynamics of Internal Layers and Diffusive Interfaces*, CBMS-NSF Regional Conf. Ser. in Appl. Math., SIAM, 1988.

91. R. A. Fisher, *The wave of advance of advantageous genes,* Ann. Eugenics 7 (1937), 355–369.

92. W. H. Fleming and H. M. Soner, *Controlled Markov Processes and Viscosity Solutions*, Applications of Mathematics, 25, Springer-Verlag, 1993.

93. S. Focant and Th. Galley, *Existence and stability of propagating fronts for an autocatalytic reaction-diffusion system*, Univ. de Paris-Sud preprint, 97–38, 1997.

94. M. Freidlin, *Propagation of a concentration wave in the presence of random motion associated with the growth of a substance*, Soviet Math. Dokl. 20 (1979), 503–507.
95. M. Freidlin, *Limit theorems for large deviations and reaction-diffusion equations*. Annals Prob. 13 (1985), 639–675.
96. M. Freidlin, *Functional Integration and Partial Differential Equations*. Annals of Mathematics Studies 109, Princeton University Press, 1985.
97. G. Frejacques, *Traveling waves in infinite cylinders with time-periodic coefficients*, Ph.D Thesis, Université Aix-Marseille, 2006.
98. Th. Gallay, *Local Stability of critical fronts in nonlinear parabolic pde's*, Nonlinearity 7 (1994), 741–764.
99. R. Gardner, *Existence of multidimensional traveling wave solutions of an initial boundary value problem*, J. Diff. Eqn. 61 (1986), 335–379.
100. J. Gärtner and M. I. Freidlin, *On the propagation of concentration waves in periodic and random media*, Soviet Math. Dokl. 20 (1979), 1282–1286.
101. J. Gärtner, W. König, S. Molchanov, *Almost sure asymptotics for the continuous parabolic Anderson model*, Probab. Theory Relat. Fields, 118 (2000), 547–573.
102. D. Gilbarg, N. Trudinger, *Elliptic Partial Differential Equations of Second Order*, Springer-Verlag, 2nd edition, 1983.
103. H. Goldstein, *Classical Mechanics*, Addison-Wesley Publishing Co, 1950.
104. D. Gomes, M. Oberman, *Computing the Effective Hamitonian Using a Variational Approach*, SIAM J. Control. Optim. 43(3) (2004), 792–812.
105. P. Groeneboom, *Brownian Motion with a Parabolic Drift and Airy Functions*, Prob. Th. Rel. Fields 81 (1989), 79–109.
106. J-S Guo, F. Hamel, *Front propagation for discrete periodic monostable equations*, Math. Ann, 335 (2006), 489–525.
107. K. P. Hadeler and F. Rothe, *Traveling fronts in nonlinear diffusion equations*, J. Math. Biology 2 (1975), 251–263.
108. T. Halpin-Healy, Y.C. Zhang, *Kinetic roughening phenomena, stochastic growth, directed polymers and all that—Aspects of multidisciplinary statistical mechanics*, Phys. Report 254 (1995), 215–415.
109. F. Hamel, *Formules min-max pour les vitesses d'ondes progressives multidimensionnelles*, Ann. Fac. Sci. Toulouse 8 (1999), 259–280.
110. F. Hamel, L. Ryzhik, *Non-adiabatic KPP fronts with arbitrary Lewis number*, Nonlinearity 18 (2005), 2881–2902.
111. F. Hamel, L. Ryzhik, *Traveling fronts for the thermal-diffusive system with arbitrary Lewis numbers*, Comm. PDE, 2009, to appear.
112. B. G. Haslam and P. D. Ronney, *Fractal properties of propagating fronts in a strongly stirred fluid*, Phys. Fluids 7 (1995), 1931–1937.
113. S. Heinze, *Wave solutions to reaction-diffusion systems in perforated domains*, Z. Anal. Anwendungen 20 (2001), 661–670.
114. S. Heinze, *The Speed of Travelling Waves for Convective Reaction-Diffusion Equations*, Preprint 84, Max-Planck-Institut für Mathematik in den Naturwissenschaften, Leipzig, Germany, 2001.
115. S. Heinze, *Large Convection Limits for KPP Fronts*, preprint 21, Max-Planck-Institut für Mathematik in den Naturwissenschaften, Leipzig, Germany, 2005.
116. S. Heinze, G. Papanicolaou, A. Stevens, *Variational principles for propagation speeds in inhomogeneous media*, SIAM J. Applied Math. 62(1), (2001), 129–148.
117. P. Hess, *Periodic–Parabolic Boundary Value Problems and Positivity*, vol. 247 of Pitman Research Notes in Mathematics, Longman Scientific & Technical, Harlow, 1991.
118. H. Holden and N. H. Risbero, *Stochastic Properties of the Scalar Buckley-Leverett Equation*, SIAM J. Appl. Math.. 51(5) (1991), 1472–1488.
119. D. Horvath, V. Petrov, S. Scott, and K. Schowalter, *Instabilities in propagating reaction-diffusion fronts*, J. Chem. Phys. 98 (1993), 6332–6343.

120. J. Huang, W. Shen, *Speeds of Spread and Propagation for KPP Models in Time Almost and Space Periodic Media*, preprint, 2008.

121. A. M. Ilin, and O. A. Oleinik, *Behavior of the solution of the Cauchy problem for certain quasilinear equations for unbounded increase of the time*, AMS Transl. (2) 42 (1964), 19–23.

122. C. Jones, *Stability of traveling wave solutions of the FitzHugh–Nagumo system*, Trans. AMS 286 (1984), 431–469.

123. G. Joulin and P. Clavin, *Linear stability analysis of nonadiabatic flames: diffusion-thermal model*, Combust. and Flame 35 (1979), 139–153.

124. J. Kanel, *Stabilization of solutions of the Cauchy problem for equations encountered in combustion theory*, Mat. Sbornik 59 (1962), 245–288.

125. I. Karatzas, S. Shreve, *Brownian Motion and Stochastic Calculus*, Springer-Verlag, 1991.

126. A. R. Kerstein, W. T. Ashurst, and F. A. Williams, *Field equations for interface propagation in an unsteady homogeneous flow field*, Phys. Rev. A 37 (1988), 2728.

127. A. R. Kerstein and W. T. Ashurst, *Propagation rate of growing interfaces in stirred fluids*, Phys. Rev. Lett. 68 (1992), 934.

128. B. Khouider, A. Bourlioux, and A. Majda, *Parameterizing turbulent flame speed-Part I: Unsteady shears, flame residence time and bending*, Combust. Theory Model. 5 (2001), 295–318.

129. I. Kim, *Homogenization of free boundary velocities*, Arch. Ration. Mech. Anal. 185 (2007), 69–103.

130. I. Kim, A. Mellet, *Homogenization of a Hele-Shaw type problem in periodic and random media*, to appear in Arch. Ration. Mech. Anal.

131. K. Kirchgässner, *On the nonlinear dynamics of traveling fronts*, J. Diff. Eqs. 96 (1992), 256–278.

132. A. Kiselev and L. Ryzhik, *Enhancement of the traveling front speeds in reaction-diffusion equations with advection*, Ann. Inst. H. Poincaré Anal. Non Linéaire 18 (2001), 309–358.

133. P. Kloeden, E. Platen, *Numerical Solution of Stochastic Differential Equations*, Springer-Verlag, Berlin, 1992.

134. A. N. Kolmogorov, I. G. Petrovsky and N. S. Piskunov, *Etude de l'eqúation de la diffusion avec croissance de la quantité de matière et son application à un problème biologique*, Moskow Univ. Math. Bull. 1 (1937), 1–25.

135. E. Kosygina, F. Rezakhanlou, and S.R.S. Varadhan, *Stochastic homogenization of Hamilton-Jacobi-Bellman equations*. Comm. Pure Appl. Math. 59(10) (2006), 1489–1521.

136. E. Kosygina and S. R. S. Varadhan, *Homogenization of Hamilton-Jacobi-Bellman equations with respect to time-space shifts in a stationary ergodic medium*, Comm. Pure Appl. Math, 61(6) (2008), 816–847.

137. N. V. Krylov, M. V. Safonov, *A property of the solutions of parabolic equations with measurable coefficients*. (English translation) Izv. Akad. Nauk SSSR Ser. Mat., 16(1) (1981), 151–164.

138. J. Krug, H. Spohn, *Kinetic roughening of growing surfaces*, In: Solids Far Away from Equilibrium, C. Godreche ed, Cambridge University Press, 1991, pp. 479–588.

139. L. D. Landau, E. M. Lifschitz, *The Classical Theory of Fields*, Oxford, Pergamon, 1960.

140. J. S. Langer, H. Müller-Krumbhaar, *Mode selection in a dendritelike nonlinear systems*, Phys. Rev. A 27 (1983), 499.

141. P. D. Lax, *Hyperbolic systems of conservation laws and the mathematical theory of shock waves*, Philadelphia SIAM Regional Conf. Ser. in Applied Math., No. 11 (1973).

142. J. Leach, D. Needham, *Matched Asymptotic Expansions in Reaction-Diffusion Theory*, Springer Monographs in Mathematics, 2003.

143. M. R. Leadbetter, G. Lindgren, H. Rootzn, *Extremes and related properties of random sequences and processes*, Springer Series in Statistics, New York, Springer-Verlag, 1983.

144. T.-Y. Lee and F. Torcaso, *Wave propagation in a lattice KPP equation in random media*, Ann. Probab. 26 (1998), 1179–1197.

145. C. Li, *Monotonicity and symmetry on bounded and unbounded domains*, I & II, Comm. PDE, 16(2-3) (1991), 491–526; 16(4-5) (1991), 585–615.

146. T. M. Liggett, *An improved subadditive ergodic theorem*, Annals of Probability 13 (1985), 1279–1285.

147. P.-L. Lions, *Generalized Solutions of Hamilton-Jacobi equations*, Pitman Research Notes in Mathematics, No. 69, Pitman Advanced Publishing Program, Boston, 1982.

148. P.-L. Lions, G. C. Papanicolaou and S. R. S. Varadhan, *Homogenization of Hamilton–Jacobi equations*, unpublished preprint, circa 1986.

149. P.-L. Lions and P. E. Souganidis, *Correctors for the homogenization of Hamilton-Jacobi equations in the stationary ergodic setting*, Comm. Pure Appl. Math. 56(10) (2003), 1501–1524.

150. P.-L. Lions and P. E. Souganidis, *Homogenization of viscous Hamilton-Jacobi equations in stationary ergodic media*, Comm. Partial Diff. Eqn. 30 (2005), 335–375.

151. P.-L. Lions and P. E. Souganidis, *Homogenization of degenerate second order PDE in periodic and almost periodic environments and applications*, Ann Inst. H. Poincaré, Analyse Non Linéaire 22 (2005), 667–677.

152. A. Majda and P. E. Souganidis, *Large scale front dynamics for turbulent reaction-diffusion equations with separated velocity scales*, Nonlinearity 7 (1994), 1–30.

153. A. Majda and P. E. Souganidis, *Flame fronts in a turbulent combustion model with fractal velocity fields*, Comm. Pure Appl. Math. LI (1998), 1337–1348.

154. A. Malevanets, A. Careta and R. Kapral, *Biscale chaos in propagating front*, Phys. Rev. E 52 (1995), 4724.

155. M. Marion, *Qualitative properties of a nonlinear system for laminar flames without ignition temperature*, Nonlinear Anal., TMA, 11 (1983), 1269–1292.

156. G. Markstein, *Nonsteady Flame Propagation*, Pergamon Press, Oxford, 1964.

157. J. E. Marsden, *Lectures on Mechanics*, London Math Soc Lecture Notes, 174, Cambridge University Press, Cambridge, 1992.

158. G. de Marsily, *Quantitative Hydrogeology—Groundwater Hydrology for Engineers*, Academic Press, 1986.

159. H. Matano, *Traveling waves in spatially inhomogeneous diffusive media*, a talk at Inst. H. Poincaré, 2002, among other conference presentations.

160. H. McKean, *Application of Brownian motion to the equation of KPP*, Comm. Pure Applied Math. 28 (1975), 323–331.

161. R. McLaughlin, J. Zhu, *The Effect of Finite Front Thickness on the Enhanced Speed of Propagation*, Combus. Sci. Tech., 129 (1997), 89–112.

162. A. Mellet, J. Nolen, L. Ryzhik, J.-M. Roquejoffre, *Stability of Generalized Transition Fronts*, Comm. in PDE, to appear.

163. A. Mellet, J.-M. Roquejoffre, *Construction d'ondes généralisé pour les models 1-d scalaire á température d'ignition*, preprint, 2008.

164. M. Metcalf, J. Merkin and S. Scott, *Oscillating wave fronts in isothermal chemical systems with arbitrary powers of autocatalysis*, Proc. Roy. Soc. Lond. A 447 (1994), 155–174.

165. J. Mierczynski, W. Shen, *Exponential separation and principal Lyapunov exponent/spectrum for random/nonautonomous parabolic equations*, J. Differential Equations 191 (2003), 175–205.

166. C. Mueller, R. Sowers, *Random Traveling Waves for the KPP equation with Noise*, J. Functional Analysis 128 (1995), 439–498.

167. J. D. Murray, *Mathematical Biology*, 2nd ed., Biomath texts, 19, 1993.

168. G. Nagin, *Équations de réaction-diffusion et propagation en milieu hétérogène*, These de Doctorat, Université Paris VI- Pierre et Marie Curie, June 2008.

169. J. Nash, *Continuity of solutions of parabolic and elliptic equations*, Amer. J. Math. 80 (1958), 931–954.

170. S. Nii, *Stability of the traveling multiple-front (multiple-back) wave solutions of the FitzHugh–Nagumo equations*, SIAM J. Math. Anal. 28 (1997), 1094–1112.

171. Y. Nishiura, Y. Oyama, K. Ueda, *Dynamics of traveling pulses in heterogeneous media of jump type*, Hokkaido Math Journal 16(1) (2007), 207–242.

172. J. Nolen, *An Invariance principle for random traveling waves in one dimension*, preprint, 2009.

173. J. Nolen, M. Rudd, J. Xin, *Existence of KPP fronts in spatially-temporally periodic advection and variational principle for propagation speeds*, Dynamics of PDE 2(1) (2005), 1–24.

174. J. Nolen, L. Ryzhik, *Traveling waves in a one-dimensional random medium*, Ann Inst. H. Poincaré, Analyse Non Linéaire, to appear, 2009.

175. J. Nolen, J. Xin *Variational Principle Based Computation of KPP Average Front Speeds in Random Shear Flows*, Methods and Applications of Analysis 11(3) (2004), 389–398.

176. J. Nolen, J. Xin, *Min-Max Variational Principle and Front Speeds in Random Shear Flows*, Methods and Applications of Analysis 11(4) (2004), 635–644.

177. J. Nolen, J. Xin, *Reaction-diffusion front speeds in spatially-temporally periodic shear flows*, SIAM J. Multiscale Modeling and Simulation 1(4) (2003), 554–570.

178. J. Nolen, J. Xin, *A Variational Principle Based Study of KPP Minimal Front Speeds in Random Shears*, Nonlinearity 18 (2005), 1655–1675.

179. J. Nolen, J. Xin, *Variational principle of KPP front speeds in temporally random shear flows with applications*, Comm. Math. Phys. 269 (2007), 493–532.

180. J. Nolen, J. Xin, *Computing Reactive Front Speeds in Random Flows by Variational Principle*, Physica D 237 (2008), 3172–3177.

181. J. Nolen, J. Xin, *KPP Fronts in a One Dimensional Random Drift*, Discrete and Continuous Dynamical Systems-B 11(2) (2009), 421–442.

182. J. Nolen, J. Xin, *Asymptotic Spreading of KPP Reactive Fronts in Incompressible Space-Time Random Flows*, Ann Inst. H. Poincaré, Analyse Non Linéaire, to appear, 2009.

183. A. Novikov, G. Papanicolaou and L. Ryzhik, *Boundary layers for cellular flows at high Peclet numbers*, Comm. Pure and Appl. Mathematics LXIII, (2005), 867–922.

184. A. Novikov, L. Ryzhik, *Bounds on the speed of propagation of the KPP fronts in a cellular flow*, Arch. Rational Mech. Anal. 184 (2007), 23–48.

185. O. A. Oleinik, *Discontinuous solutions of the nonlinear differential equations*, Uspehi Mat. Nauk (N.S.) 12(3) (1957) 3–73; AMS Translations, pp. 95–172, 1963.

186. H. Osada, *Diffusion processes with generators of generalized divergence form*, J. Math. Kyoto Univ 27(4)(1987), 597–619.

187. S. Osher, R. Fedkiw, *Level Set Methods and Dynamic Implicit Surfaces*, Applied Math Sci, 153, Springer, New York, 2003.

188. S. Osher and J. Ralston, *Stability of traveling waves in convective porous medium equation*, Comm. Pure Appl. Math. 35 (1982), 737–749.

189. G. Papanicolaou, *Diffusion in Random Media*, Surveys in Applied Mathematics, edited by J. B. Keller, D. McLaughlin, and G. Papanicolaou, Plenum Press, (1995), pp. 205–255.

190. G. Papanicolaou and S. Varadhan, *Boundary value problems with rapidly oscillating random coefficients*, Colloq. Math. Soc. Janos Bolyai 27, Random Fields, Esztergom, Hungary (1979), Amsterdam, North Holland, 1982, pp. 835–873.

191. G. Papanicolaou and J. Xin, *Reaction-diffusion fronts in periodically layered media*, J. Stat. Physics 63(5/6) (1991), 915–931.

192. K. Park, I. M. Kim, *Dynamics of an interface driven through random media: The effect of spatially correlated noise*, J. Phys. Soc. Japan 72 (2003), 111–116.

193. N. Peters, *A spectral closure for premixed turbulent combustion in the flamelet regime*, J. Fluid Mech. 242 (1992), 611–629.
194. N. Peters, *Turbulent Combustion*, Cambridge University Press, Cambridge, 2000.
195. J. R. Phillip, *Theory of Infiltration*, Adv. in Hydrosciences 5 (1969), 213–305.
196. J. R. Phillip, *Issues in Flow and Transport in Heterogeneous Porous Media*, Transport in Porous Media 1 (1986), 319–338.
197. M. Postel, J. Xin, *A numerical study of fronts in random media using a reactive solute transport model*, Computational Geoscience 1 (1997), 251–270.
198. M. Protter and H. Weinberger, *Maximum Principles in Differential Equations*, Prentice-Hall, 1967.
199. J. Qian, *Two approximations for Effective Hamiltonians Arising from Homogenization of Hamilton-Jacobi Equations*, CAM report 03-39, UCLA, August, 2003.
200. F. Rezakhanlou, J. Tarver, *Homogenization for Stochastic Hamilton-Jacobi Equations*, Arch. Rational Mech. Anal, 151 (2000), 277–309.
201. F. Rezakhanlou, *Central Limit Theorem for Stochastic Hamilton-Jacobi Eqautions*, Comm. Math. Phys. 211 (2000), 413–438.
202. L. C. Rogers, D. Williams, *Diffusion, Markov Processes and Martingales*, Vol. 1, Foundations, John Wiley and Sons, 1994.
203. P. Ronney, *Some Open Issues in Premixed Turbulent Combustion*, Lecture Notes in Physics, 449, Springer-Verlag, Berlin, 1995, pp. 1–22.
204. P. D. Ronney, B. G. Haslam, and N. O. Rhys, *Front propogation rates in randomly stirred media*, Phys. Rev. Lett. 74 (1995), 3804–3807.
205. J.-M. Roquejoffre, *Eventual monotonicity and convergence to traveling fronts for the solutions of parabolic equations in cylinders*, Ann Inst. H. Poincaré, Analyse Nonlineare 14 (1997), 499–552.
206. S. Ross, *Introduction to Probability Models*, 9th edition, Academic Press, 2007.
207. B. L. Rozdestvenskii, *The Cauchy Problem for Quasilinear Equations*, Dokl. Akad. Nauk SSSR 115 (1957), pp. 454–457; AMS Translations, Series 2 42 (1964), 25–30.
208. L. Ryzhik and A. Zlatoš, *KPP pulsating front speed-up by flows*, Comm. Math Sci 5 (2007), 575–593.
209. H. Rund, *The Hamilton-Jacobi theory in the calculus of variations: its role in mathematics and physics*, Van Nostrand Co, London, 1966.
210. Wm. van Saarloos, *Front propagation into unstable states*, Physics Reports 386 (2003), 29–222.
211. B. Sandstede, *Stability of N-fronts bifurcating from a twisted heteroclinic loop and an application to the FitzHugh–Nagumo equation*, SIAM. J. Math. Anal. 29 (1998), 183–207.
212. D. H. Sattinger, *On the stability of waves of nonlinear parabolic systems*, Advances in Math. 22 (1976), 312–355.
213. R. Schwab, *Stochastic Homogenization of Hamilton-Jacobi Equations in Stationary Ergodic Spatial-Temporal Media*, preprint (2008), Indiana Univ Math J, to appear.
214. S. Schumacher, *Diffusions with random coefficients*, in: Particle Systems, Random Media and Large Deviations (R. Durrett Ed.), American Mathematical Society, Providence: 1985.
215. L. Shen, J. Xin, A. Zhou, *Finite Element Computation of KPP Front Speeds in Random Shear Flows in Cylinders*, SIAM J. Multiscale Modeling and Simulation 7(3) (2008), 1029–1041.
216. W. Shen, *Traveling waves in time almost periodic structures governed by bistable nonlinearities, I. stability and uniqueness*, J. Diff. Eqn. 159 (1999), 1–55.
217. W. Shen, *Traveling waves in time almost periodic structures governed by bistable nonlinearities, II. existence*, J. Diff. Eqn. 159 (1999), 55–101.
218. W. Shen, *Traveling Waves in Diffusive Random Media*, J. Dynamics Diff. Eqs. 16(4) (2004), 1011–1060.

219. W. Shen, *Traveling waves in time dependent bistable media*, Diff Int. Eqn. 19(3) (2006), 241–278.
220. W. Shen, *Variational Principle for Spatial Spreading Speeds and Generalized Wave Solutions in Time Almost Periodic and Space Periodic KPP Models*, preprint, 2008.
221. N. Shigesada, K. Kawasaki and E. Teramoto, *Traveling periodic waves in heterogeneous environments*, Theor. Population Biology 30 (1986), 143–160.
222. S. Shy, R. Jang, P. Ronney, *Laboratory Simulation of Flamelet and Distributed Models for Premixed Turbulent Combustion Using Aqueous Autocatalytic Reactions*, Combustion Science and Technology, 113-114 (1996), 329–340.
223. G. I. Sivashinsky, *Instabilities, pattern formation, and turbulence in flames*, Ann. Rev. Fluid Mech. 15 (1983), 179–199.
224. G. I. Sivashinsky, *Cascade-renormalization theory of turbulent flame speed*, Combust. Sci. Technology 62 (1988), 77–96.
225. M. E. Smaily, *Pulsating traveling fronts: Asymptotics and homogenization regimes,* European J. Appl. Math, to appear.
226. J. Smoller, *Shock Waves and Reaction-Diffusion Equations*, Graduate Text in Mathematics, Vol. 258, Springer-Verlag, 1983.
227. P. E. Souganidis, *Stochastic homogenization of Hamilton-Jacobi equations and some applications*, Asymptotic Analysis 20 (1999), 1–11.
228. K. Uchiyama, *The behavior of some nonlinear diffusion equations for large time*, J. Math. Kyoto Univ. 18 (1978), 453–508.
229. S. R. S. Varadhan, *Large Deviations and Applications*, CBMS-NSF Regional Conference Series in Applied Math., 46, SIAM, 1984.
230. S. R. S. Varadhan, N. Zygouras, *Behavior of the solution of a random semilinear heat equation*, Comm. Pure Appl. Math. 61 (2008), 1298–1329.
231. N. Vladimirova, P. Constantin, A. Kiselev, O. Ruchayskiy, L. Ryzhik, *Flame enhancement and quenching in fluid flows*, Combust. Theory Model. 7 (2003), 487–508.
232. A. I. Volpert, V. A. Volpert, *Traveling wave solutions of parabolic systems*, Translations of Math Monographs, 140, American Math Soc, 1994.
233. J. Wehr, J. Xin, *White Noise Perturbation of the Viscous Shock Fronts of the Burgers Equation*, Commun. Math. Phys. 181 (1996), 183–203.
234. J. Wehr, J. Xin, *Front speed in a Burgers equation with random flux*, J. Stat. Phys. 88(3/4) (1997), 843–871 .
235. J. Wehr, J. Xin, *Scaling Limits of Waves in Convex Scalar Conservation Laws under Random Initial Perturbations*, J. Stat. Physics 122(2) (2006), 361–370.
236. H. Weinberger, *On spreading speed and traveling waves for growth and migration models in a periodic habitat*, J. Math. Biol 45 (2002), 511–548.
237. G. B. Whitham, *Linear and Nonlinear Waves*, Wiley and Sons, 1979.
238. F. Williams, *Combustion Theory*, Benjamin Cummings, Menlo Park, CA, second edition, 1985.
239. W. Woyczyński, *Burgers-KPZ Turbulence–Göttingen Lectures*, Lecture Notes in Mathematics, 1700, Springer, 1998.
240. J. Xin, *Existence and stability of travelling waves in periodic media governed by a bistable nonlinearity*, J. Dynamics Diffl. Eqs. 3 (1991), 541–573.
241. J. Xin, *Existence and uniqueness of travelling wave solutions in a reaction-diffusion equation with combustion nonlinearity*, Indiana Math. J. 40 (1991), 985–1008.
242. J. Xin, *Existence of planar flame fronts in convective-diffusive periodic media*, Arch. Rat. Mech. Anal. 121 (1992), 205–233.
243. J. Xin, *Existence and nonexistence of traveling waves and reaction-diffusion front propagation in periodic media*, J. Stat. Phys. 73 (1993), 893–926.
244. J. Xin, *Existence of multidimensional traveling waves in transport of reactive solutes through periodic porous media*, Arch. Rat. Mech. Anal. 128 (1994), 75–103.
245. J. Xin, *Stability of traveling waves in a solute transport equation*, J. Diffl. Eqs., 135 (1997), 269–298.

246. J. Xin, *Front propagation in heterogeneous media*, SIAM Review, 42(2) (2000), 161–230.

247. J. Xin, *KPP front speeds in random shears and the parabolic Anderson problem*, Methods and Applications of Analysis, 10(2) (2003), 191–198.

248. J. Xin, A. Peirce, J. Chadam, P. Ortoleva, *Reactive flows in layered porous media, I. Homogenization of free boundary problems*, Asymptotic Analysis 11 (1995), 31–54.

249. J. Xin, A. Peirce, J. Chadam, P. Ortoleva, *Reactive flows in layered porous media, II. The Shape Stability of the Reaction Interface*, SIAM J. Applied Math, 53/2 (1993), 319–339.

250. J. Xin, J. Zhu, *Quenching and propagation of bistable reaction-diffusion fronts in multidimensional periodic media*, Physica D 81 (1995), 94–110.

251. V. Yakhot, *Propagation velocity of premixed turbulent flames*, Comb. Sci. Tech. 60 (1988), 191–241.

252. X. Yuan, T. Teramoto, Y. Nishiura, *Heterogeneity-induced defect bifurcation and pulse dynamics for a three-component reaction-diffusion system*, Physical Review E, 75, 036220 (2007), 1–12.

253. S. van der Zee and W. H. van Riemsdijk, *Transport of reactive solute in spatially variable soil systems*, Water Resources Research 23 (1987), 2059–2069.

254. E. Zeidler, *Nonlinear Functional Analysis*, Springer-Verlag, 1984.

255. A. Zlatoš, *Pulsating front speed-up and quenching of reaction by fast advection*, Nonlinearity 20 (2007), 2907–2921.

256. A. Zlatoš, *Sharp asymptotics for KPP pulsating front speed-up and diffusion enhancement by flows*, to appear in Arch. Ration. Mech. Anal, 2009.

Index

151882LV00003B/31/P
Printed in the United States